CONTENTS

GRADE 4

ISBN: 978-1-927042-13-7

1 Numbers to 10 000

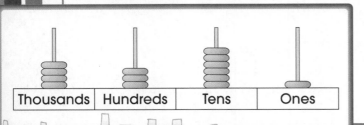

Thousands	Hundreds	Tens	Ones

Standard form: 4361

Expanded form: 4000 + 300 + 60 + 1

Written form: Four thousand three hundred sixty-one

How many eggs are there in the trucks? Write the numbers in different forms.

1111 = 1000 + 100 + 10 + 1
One thousand one hundred eleven

①

Standard form: _3456_

Expanded form: _3000+400+50+6_

Written form: _three thousand four hundred fifty-six_

②

Standard form: _4235_

Expanded form: _4000+200+30+5_

Written form: _four thousand two hundred thirty-five_

See how many eggs were sold last week. Write the numbers in words.

Number of Eggs

③ SUN: 4005 — four thousand and five

④ MON: 3200 — three thousand two hundred

⑤ TUE: 1060 — one thousand sixty

⑥ WED: 5274 — five thousand two hundred seventy four

⑦ THU: 9863 — nine thousand eight hundred sixty three

⑧ FRI: 7049 — seven thousand fourty - nine

⑨ SAT: 6751 — six thousand seven hundred fifty - one

Write the numbers in standard form.

⑩ 5 thousands   2 hundreds   7 tens   6 ones — 5276

⑪ 3 thousands   4 hundreds   8 tens   2 ones — 3482

⑫ 2 thousands   3 hundreds   9 ones — 2309

⑬ 7 thousands   5 hundreds   4 tens — 7540

Write the value of each coloured digit.

⑭ 4369 — 300

⑮ 2501 — 2000

⑯ 6887 — 80

⑰ 1009 — 9

Thousands   Tens
| Hundreds | Ones
↓   ↓   ↓   ↓
9   5   4   6

9 is in the thousands column. It means 9000.

ISBN: 978-1-927042-13-7

3294   3924
If the digits in the thousands column are the same, compare the digits in the hundreds column and so on.

Compare the numbers. Put ">" or "<" in each circle.

⑱ 2088 $<$ 3198

⑲ 5129 $<$ 5912

⑳ 6104 $>$ 6091

㉑ 1033 $<$ 1036

㉒ 7365 $>$ 7356

> : greater than

< : smaller than

3294 < 3924

Look at the numbers on the flowers. Put each group of numbers in order. Start with the greatest.

㉓ 2308 4802 1988 2867

> 4802, 2867, 2308, 1988

㉔ 3264 4632 6342 2364

6342, 4632, 3264, 2364.

㉕ 5070 5700 5007 7005

7005, 5700, 5070, 5007

Fill in the missing numbers.

㉖ 6700 _6600_ 6500 _6400_ _6300_ 6200

㉗ 8020 8030 _8040_ _8050_ 8060 _8070_

㉘ 9444 8444 _7444_ 6444 _5444_ _4444_

㉙ 5099 _6000_ 6001 _6002_ _6003_ 6004

Help Wayne form the smallest 4-digit number by using the given digits in each group. Then fill in the blanks.

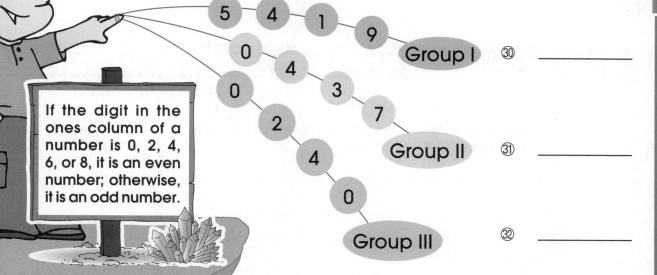

If the digit in the ones column of a number is 0, 2, 4, 6, or 8, it is an even number; otherwise, it is an odd number.

Group I ㉚ _____

Group II ㉛ _____

Group III ㉜ _____

㉝ Group _____ forms the smallest 4-digit number.

㉞ The number formed by Group _____ has 4 in its hundreds column.

㉟ The number formed by Group _____ is an even number.

㊱ Write all the possible 4-digit numbers formed by Group III. Put them in order from smallest to greatest.

_____

ACTIVITY

Follow the pattern to fill in the missing numbers.

① 101 x 44 = _____

② 101 x 55 = _____

③ 101 x _____ = 6666

④ 101 x _____ = 7777

$$101 \times 11 = 1111$$
$$\downarrow \text{times 2} \quad \downarrow \text{times 2}$$
$$101 \times 22 = 2222$$

$$101 \times 11 = 1111$$
$$\downarrow \text{times 3} \quad \downarrow \text{times 3}$$
$$101 \times 33 = 3333$$

2 Addition and Subtraction

Estimate – guess how many or how much

Round – express a number to the nearest whole number, ten, or other values

Round 4652 to the nearest hundred.

Look at the digit in the tens column.

$4652 \longrightarrow 4700$

Remember to add the digit carried over from the right.

$$
\begin{array}{r}
\;\overset{①}{3\,0\,0\,9} \\
+\;2\,2\,6\,3 \\
\hline
5\,2\,7\,2
\end{array}
$$

Do the addition. Follow the sums to plot Rene's way home. Draw the lines.

To do vertical addition, remember to align the numbers on the right.

$$
\begin{array}{r}
3\,2\,1\,9 \\
+\;\;\;1\,5\,7
\end{array}
$$

①
$$
\begin{array}{r}
1\,2\,3\,4 \\
+\;\;\;2\,5\,4 \\
\hline
1488
\end{array}
$$

②
$$
\begin{array}{r}
3\,2\,6\,9 \\
+\;4\,8\,4\,3 \\
\hline
8112
\end{array}
$$

③
$$
\begin{array}{r}
5\,0\,7\,5 \\
+\;\;\;9\,3\,5 \\
\hline
6010
\end{array}
$$

④
$$
\begin{array}{r}
4\,6\,2\,7 \\
+\;2\,1\,3\,2 \\
\hline
6759
\end{array}
$$

⑤
$$
\begin{array}{r}
5\,6\,6 \\
+\;8\,7\,7\,7 \\
\hline
9343
\end{array}
$$

⑥
$$
\begin{array}{r}
2\,5\,9\,9 \\
+\;4\,7\,8\,6 \\
\hline
7385
\end{array}
$$

⑦ 6586 + 987 = 7573

⑧ 7582 + 1429 = _____

⑨ 4999 + 1655 = _____

⑩ 4555 + 2187 = _____

⑪

8112 •   • 7385

6010 •

1488 •

• 7573

6742 •

6759 •   9343 •

9011 •   • 6654

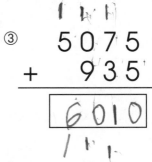
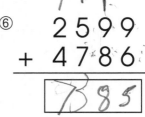
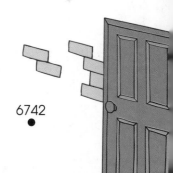
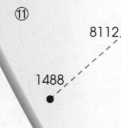
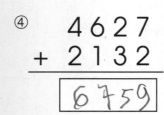

Do the subtraction.

⑫
```
  5 9 6 7
-   1 5 6
```

⑬
```
  9 6 9 9
-   2 4 7
```

⑭
```
  2 4 2 3
-   3 1 3
```

⑮
```
  4 5 8 2
-   4 5 9
```

⑯
```
  3 0 6 3
-   7 2 5
```

⑰
```
  8 2 4 9
-   9 7 3
```

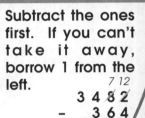

Subtract the ones first. If you can't take it away, borrow 1 from the left.
```
      7 12
  3 4 8 2
-   3 6 4
```

⑱ 8761 - 601 = _____

⑲ 6109 - 218 = _____

⑳ 7000 - 815 = _____

㉑ 5222 - 777 = _____

(Check)
```
  1000        812
- 812       + 188
  188
```
If this answer is 1000, it means the answer (188) you get from the subtraction is correct.

Check the answer to each question. Put a check mark ✔ in the circle if it is correct. Put a cross ✗ if it is not.

(Check)

㉒
```
  3 0 0 0
-   8 6 5
  2 1 3 5  ◯
```

(Check)

㉓
```
  5 1 2 5
-   3 2 9
  4 6 9 6  ◯
```

(Check)

㉔
```
  8 1 2 1
-   9 7 7
  7 0 4 4  ◯
```

(Check)

㉕
```
  7 0 4 6
-   6 8 3
  6 3 6 3  ◯
```

ISBN: 978-1-927042-13-7

Round to the nearest thousand:

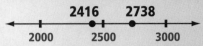

2416, being closer to 2000 than 3000, is rounded to 2000.

2738, being closer to 3000 than 2000, is rounded to 3000.

2500, being halfway between 2000 and 3000, is rounded to 3000.

A number halfway between two numbers is rounded to the greater number, e.g.

$$1500 \longrightarrow 2000$$

Round the numbers to the nearest hundred.

㉖ 1462 _____   ㉗ 6819 _____

㉘ 5203 _____   ㉙ 1067 _____

㉚ 7134 _____   ㉛ 9872 _____

㉜ 3090 _____   ㉝ 1174 _____

Round the numbers to the nearest thousand.

㉞ 8864 _____   ㉟ 3217 _____

㊱ 3551 _____   ㊲ 4279 _____

㊳ 9250 _____   ㊴ 8043 _____

Help Susan do the estimates by rounding the numbers to the nearest hundred. Then find the exact answers.

㊵
```
   4 2 1 6
 + 3 7 0 4
```
Estimate

㊶
```
   8 9 5 0
 -   8 9 9
```
Estimate

㊷
```
   7 0 9 5
 -   9 1 9
```
Estimate

㊸
```
   5 6 1 6
 + 1 8 6 4
```
Estimate

㊹ 9099 − 670 = _____   Estimate _____

㊺ 3500 + 2651 = _____   Estimate _____

Answer the questions.

㊻ The concert hall is divided into 2 sections. If there are 1258 seats in each section, how many seats are there in the concert hall in all?

_____ = _____

There are _____ seats in the concert hall in all.

㊼ 2016 people attended the concert yesterday. If there were 986 women, how many men were there?

_____ = _____

There were _____ men.

㊽ Jackson, the singer, expects to earn half of ten thousand dollars from the concert. If he earns $817 less than expected, how much does he earn?

_____ = _____

He earns $ _____ .

Half of **10** is **5**.
Half of **100** is **50**.
Half of **1000** is **500**.

Follow this pattern to find half of ten thousand.

ACTIVITY

Write the correct digits in the boxes.

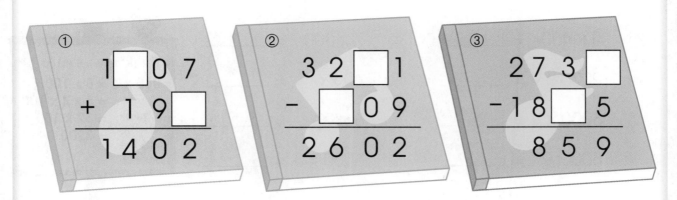

①
```
  1 □ 0 7
+   1 9 □
─────────
  1 4 0 2
```

②
```
  3 2 □ 1
-   □ 0 9
─────────
  2 6 0 2
```

③
```
  2 7 3 □
- 1 8 □ 5
─────────
    8 5 9
```

3 Multiplication

WORDS TO LEARN

Multiplication – a short way to find a sum when the addends are the same

Product – the answer you get after multiplying

$$\text{Number of stars:} \quad \underset{\text{multiplicand}}{3} \quad \times \quad \underset{\text{multiplier}}{4}$$
$$= 12 \leftarrow \text{product}$$

> When you multiply a number by 10, 100, or 1000, just add the same number of zero(es) to the number to get the answer.

Do the multiplication mentally.

① $5 \times 10 = $ _____

② $10 \times 9 = $ _____

③ $10 \times 8 = $ _____

④ $1 \times 100 = $ _____

⑤ $7 \times 100 = $ _____

⑥ $100 \times 2 = $ _____

⑦ $3 \times 1000 = $ _____

⑧ $4 \times 1000 = $ _____

Find the products.

⑨ $4 \times 200 = 4 \times $ _____ $\times 100$

$= $ _____ $\times 100$

$= $ _____

⑩ $2 \times 30 = 2 \times $ _____ $\times 10$

$= $ _____ $\times 10$

$= $ _____

⑪ $5 \times 400 = $ _____

⑫ $7 \times 300 = $ _____

⑬ $2000 \times 9 = $ _____

⑭ $40 \times 6 = $ _____

⑮ $3000 \times 4 = $ _____

⑯ $8 \times 90 = $ _____

⑰ $500 \times 9 = $ _____

⑱ $6000 \times 3 = $ _____

Think: $800 = 8 \times 100$
$800 \times 4 = 8 \times 100 \times 4$
$= 8 \times 4 \times 100$
$= 32 \times 100$
$= 3200$

A quick way to find the answer!

ISBN: 978-1-927042-13-7

2-digit number x 1-digit number

1st Multiply the ones.

```
    1
    2 6
  x   3
      8
      ↑
```
6 x 3 = 18 (Carry 1 ten to the tens column.)

2nd Multiply the tens.

```
    1
    2 6
  x   3
    7 8
    ↑
```
2 x 3 = 6; 6 + 1 = 7

Find the answers.

⑲
```
    1 3
  x   2
```

⑳
```
    2 4
  x   4
```

㉑
```
    1 8
  x   3
```

Sometimes you need to carry more than 1 ten to the tens column. 4 ← 4 tens
e.g.
```
    3 6
  x   7
  2 5 2
```

㉒
```
    3 2
  x   5
```

㉓
```
    9 8
  x   6
```

㉔
```
    7 9
  x   1
```

㉕ 6 x 21 = _____

㉖ 73 x 4 = _____

㉗ 5 x 49 = _____

㉘ 26 x 8 = _____

㉙ 83 x 2 = _____

㉚ 37 x 7 = _____

Though the order of multiplication has changed, the product is the same.
e.g. 2 x 14 = 14 x 2

$37

㉛ a. 5 helmets cost $ _____ .

b. 8 helmets cost $ _____ .

$8

㉜ a. 15 pairs of knee pads cost $ _____ .

b. 34 pairs of knee pads cost $ _____ .

ISBN: 978-1-927042-13-7

Read what Uncle Jeffrey says. Then help him write the numbers.

How many cookies are there in 4 boxes?

③③ **Multiply the ones.**

Carry 1 ten to the tens column.

$$
\begin{array}{r}
{}^{1}134 \\
\times 4 \\
\hline \square
\end{array}
$$

Multiply the tens.

Carry 1 hundred to the hundreds column.

$$
\begin{array}{r}
{}^{1}1{}^{1}34 \\
\times 4 \\
\hline \square 6
\end{array}
$$

Multiply the hundreds.

$$
\begin{array}{r}
{}^{1}134 \\
\times 4 \\
\hline \square 36
\end{array}
$$

There are _____ cookies in 4 boxes.

Find the products.

3-digit number × 1-digit number

1st Multiply the ones.
2nd Multiply the tens.
3rd Multiply the hundreds.

③④
$$
\begin{array}{r}
213 \\
\times 3 \\
\hline
\end{array}
$$

③⑤
$$
\begin{array}{r}
375 \\
\times 4 \\
\hline
\end{array}
$$

③⑥
$$
\begin{array}{r}
519 \\
\times 5 \\
\hline
\end{array}
$$

③⑦ $183 \times 6 =$ _____

③⑧ $274 \times 9 =$ _____

③⑨ $408 \times 8 =$ _____

④⓪ $812 \times 7 =$ _____

Look at the boxes of cereal. Then complete the tables.

④①
No. of Boxes	3	7	8	9
Total weight (g)				

④②
No. of Boxes	2	5	6
Total weight (g)			

Answer the questions.

43. Uncle Philip can make 18 pizzas in an hour. How many pizzas can he make in 9 hours?

    He can make _____ pizzas.

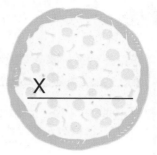

44. If 325 pizzas are sold every day, how many pizzas will be sold in a week?

    _____ pizzas will be sold.

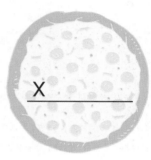

45. If Mrs. Venn buys 15 medium-sized pizzas and 8 large-sized pizzas, how much does she need to pay?

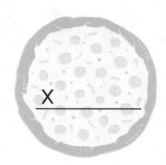 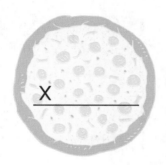 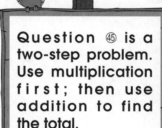

Question 45 is a two-step problem. Use multiplication first; then use addition to find the total.

Small: $7
Medium: $9
Large: $13

    She needs to pay $ _____ .

 **A C T I V I T Y**

Answer the question.

How many slices of pizza weigh about 800 g?

136 g

_____ slices of pizza

Division

WORDS TO LEARN

Division – the inverse operation of multiplication
Remainder – what is left after you divide

e.g.

dividend   quotient
divisor   remainder

$15 \div 2 = 7R1$

$$\begin{array}{r} 7\ R\ 1 \\ 2\overline{)15} \\ 14 \\ \hline 1 \end{array}$$

Read what Mrs. Lastman says. Help her write the numbers and do the division.

I pick 74 flowers and put them into groups of 6. How many groups are there? How many flowers are left?

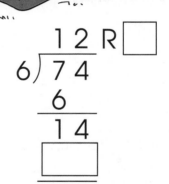

4 steps to do division:

1st Divide.
2nd Multiply.
3rd Subtract.
4th Bring down.

①

$$\begin{array}{r} 1 \\ 6\overline{)74} \\ 6 \\ \hline \square \end{array}$$ ← multiply $6 \times 1 = 6$ ← subtract

$$\begin{array}{r} 1 \\ 6\overline{)74} \\ 6 \\ \hline 1\square \end{array}$$ ↓ bring down

$$\begin{array}{r} 12\ R\ \square \\ 6\overline{)74} \\ 6 \\ \hline 14 \\ \square \\ \hline \square \end{array}$$

There are _____ groups with _____ flowers left.

② $3\overline{)69}$

③ $4\overline{)52}$

④ $7\overline{)96}$

⑤ $5\overline{)84}$

⑥ $91 \div 7 =$ _____

⑦ $77 \div 4 =$ _____

⑧ $91 \div 6 =$ _____

⑨ $82 \div 9 =$ _____

⑩ $25 \div 3 =$ _____

⑪ $43 \div 5 =$ _____

Help Patrick the Dog write the numbers.

⑫

Divide the hundreds.	Divide the tens.	Divide the ones.

Divide the hundreds:
$$3\overline{)445}$$

Divide the tens:
$$3\overline{)445}$$
3 ↓ bring down

Divide the ones:
$$3\overline{)445}$$ R
3
14
12 ↓ bring down
2

$445 \div 3 =$ _____

Do the division.

When the dividend is smaller than the divisor, put a "0" in the quotient.
$$3\overline{)312}$$ = 104
3
12
12

⑬
$$4\overline{)484}$$

⑭
$$7\overline{)882}$$

⑮
$$6\overline{)715}$$

⑯
$$9\overline{)953}$$

⑰ $636 \div 3 =$ _____

⑱ $439 \div 7 =$ _____

⑲ $504 \div 6 =$ _____

⑳ $188 \div 8 =$ _____

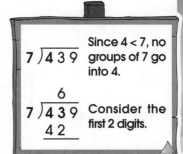

Since $4 < 7$, no groups of 7 go into 4.
$$7\overline{)439}$$

Consider the first 2 digits.
$$7\overline{)439}$$
42

㉑ Patrick divides 360 bones equally into 5 groups.

There are _____ bones in each group.

㉒ Patrick shares 119 bones with 3 friends. Each dog

gets _____ bones. _____ bones are left.

ISBN: 978-1-927042-13-7

Cross out ✗ the mistakes and write the correct division.

㉓
```
      1 5̷
   5 ) 7 2        5 ) 7 2
      5
      2 2
      2 5
```

㉔
```
      1
   3 ) 8 2        3 ) 8 2
      3
      5 2
```

㉕
```
      1 2
   3 ) 3 8        3 ) 3 8
      3
      8
      6
      1
```

㉖
```
      1 3
   6 ) 8 0        6 ) 8 0
      6
      2 0
      1 6
      4
```

Use multiplication to check each answer. Put a check mark ✔ in the box if the answer is correct. Write the correct answer if it is not.

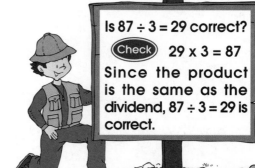

Is $87 \div 3 = 29$ correct?
Check $29 \times 3 = 87$
Since the product is the same as the dividend, $87 \div 3 = 29$ is correct.

㉗ $90 \div 5 = \underline{\ 18\ }$ ☐

Check _____

㉘ $74 \div 2 = \underline{\ 38\ }$ ☐

Check _____

㉙ $760 \div 4 = \underline{\ 180\ }$ ☐   Check _____

㉚ $861 \div 7 = \underline{\ 123\ }$ ☐   Check _____

4

Help Mrs. Wilson solve the problems.

A box of 6	🧁	.... $9
A box of 8	🍩	.... $7
A pack of 5	🥪	.... $22

㉛ If Mrs. Wilson buys 105 sandwiches, how many packs of sandwiches does she buy?

_____ = _____

_____ packs

㉜ Mrs. Wilson pays $168 for buying doughnuts. How many boxes of doughnuts does she buy?

_____ = _____

_____ boxes

㉝ If Mrs. Wilson spends $100 more on doughnuts, how many more boxes of doughnuts can she buy?

_____ = _____   _____ more

㉞ If Mrs. Wilson needs 197 muffins, how many boxes of muffins does she need to buy?

_____ = _____   _____ boxes

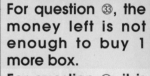

For question ㉝, the money left is not enough to buy 1 more box.
For question ㉞, it is necessary to buy 1 more box.

A C T I V I T Y

Solve the problem.

Mary has 2 boxes of cookies. If she shares her cookies with 3 friends, how many cookies will each child get?

_____ cookies

ISBN: 978-1-927042-13-7

5 Time

WORDS TO LEARN

Timeline - a schedule of activities or events
Units for time - year, decade, century, millennium

1 decade   = 10 years
1 century   = 10 decades (100 years)
1 millennium = 10 centuries (1000 years)

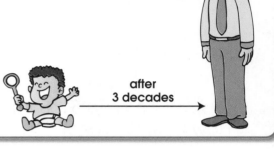

after
3 decades

The timeline shows when the things were invented. Use the information to fill in the blanks.

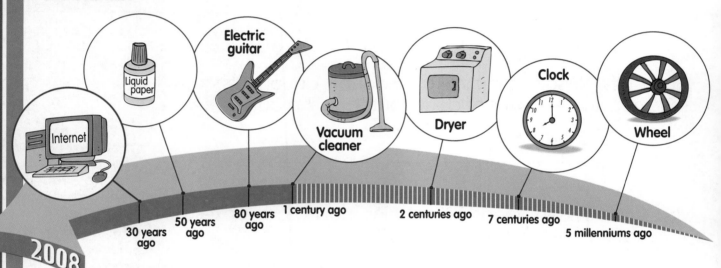

① The wheel was invented about _____ centuries ago.

② The clock was invented about _____ decades ago.

③ The _____ was invented about 3 decades ago.

④ The _____ was invented about 10 decades ago.

⑤ The _____ was invented about 8 decades ago.

⑥ The _____ was invented about 20 decades ago.

Fill in the blanks.

⑦ 5 centuries   = _____ decades   ⑧ 3 decades   = _____ years

⑨ 2 millenniums = _____ centuries   ⑩ 4 centuries   = _____ years

⑪ 80 years   = _____ decades   ⑫ 7 millenniums = _____ years

Canadian Curriculum MathSmart (Grade 4)   ISBN: 978-1-927042-13-7

Find how long it took the children to plant their saplings. Then answer the questions.

You can use subtraction to find the duration.

$$\begin{array}{ll} 9{:}47 & \text{Finishing time} \\ -\ 9{:}23 & \text{Starting time} \\ \hline 24 & \text{Duration} \end{array}$$

Tony Mike Katie Judy

	Tony	Mike	Katie	Judy
Starting time	3:07	2:25	11:09	12:28
Finishing time	3:49	2:53	11:31	12:56
⑬ **Time taken (min)**				

⑭ Who took the longest time to plant the sapling? _____

⑮ How many children took less than half an hour to plant the saplings? _____

⑯ What is the difference in the time taken between the two boys? _____

ACTIVITY

Solve the problem.

If Kevin leaves home at 7:37 and Jack leaves home at 7:34, who arrives at school first?

It takes me 24 minutes to walk to school.

It takes me 19 minutes to ride to school.

Kevin

Jack

_____

ISBN: 978-1-927042-13-7

6 Perimeter and Area

WORDS TO LEARN

Perimeter – the distance around the outside of a shape

Area – the number of square units of a surface

Units for length or distance – kilometre (km), metre (m), decimetre (dm) centimetre (cm), millimetre (mm)

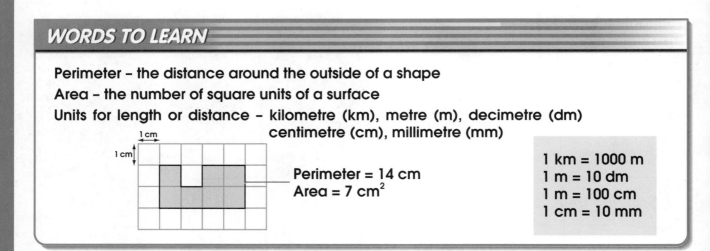

Perimeter = 14 cm
Area = 7 cm^2

1 km = 1000 m
1 m = 10 dm
1 m = 100 cm
1 cm = 10 mm

Look at the picture. Write the most appropriate unit (km, m, dm, cm, or mm) to complete each record.

①

1 _____

36 _____

104 _____

2 _____

125 _____

9 _____

62 _____

4 _____

12 _____

18 _____

Find the perimeter of each shape.

②

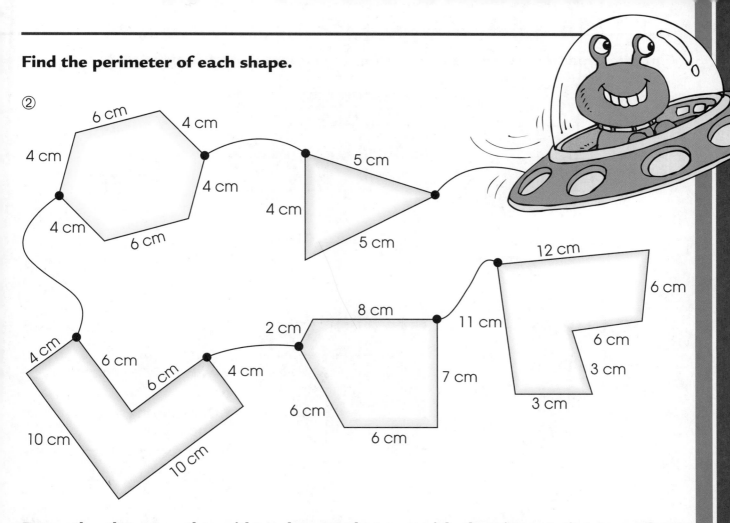

Draw the shape on the grid to show each room with the given perimeter. Then label it.

③

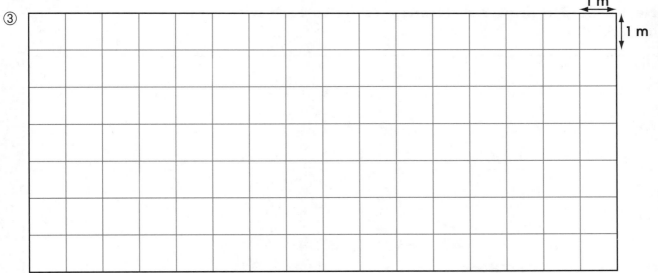

The rooms are in the shape of either a square or a rectangle.

Perimeter	
Bedroom:	18 m
Kitchen:	10 m
TV room:	12 m
Bathroom:	8 m

Find the area of each shape by counting the squares and triangles.

④

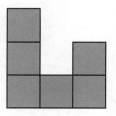

_____ cm²

⑤

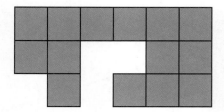

_____ cm²

⑥

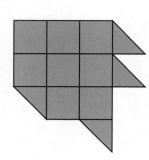

_____ cm²

⑦

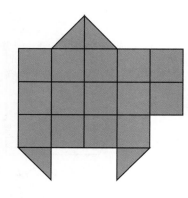

_____ cm²

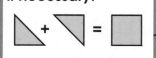

Combine the shapes if necessary.

▷ + ◺ = ◻

Help the alien draw 3 different rectangles each with an area of 12 m².

⑧

1 m²

ISBN: 978-1-927042-13-7

Help Farmer Wright find the perimeter and area of each section of his field. Then answer the questions.

A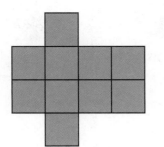

B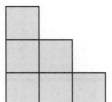

C

D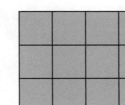

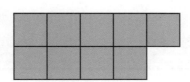

When two shapes have the same area, it does not necessarily mean that they have the same perimeter.

⑨

Section	A	B	C	D
Perimeter (units)				
Area (square units)				

⑩ Which section has the greatest area? _____

⑪ The section with the smallest perimeter is for planting tomatoes. Which section is it? _____

⑫ a. Which 2 sections have the same perimeter? _____

b. Do they have the same area? _____

  ACTIVITY

Check ✔ the shape that has the same perimeter as the one on the left.

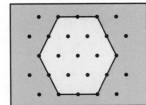 

A B C

ISBN: 978-1-927042-13-7

7 2-D Shapes

WORDS TO LEARN

Quadrilateral – a figure with 4 straight sides

 **Square**   **Rectangle**   **Parallelogram**   **Trapezoid**   **Rhombus**

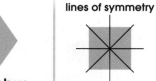
lines of symmetry

A square has 4 lines of symmetry.

Parallel lines – lines that never meet (e.g. //)

Perpendicular lines – lines that meet at a right angle (e.g. ⌐)

Congruent figures – figures that have the same shape and size (e.g. and)

Similar figures – figures that have the same shape but different sizes (e.g. and)

Angle – the space between two meeting lines

Unit for angle – Degree (°)

Colour the quadrilaterals. Then draw lines to match them with their names.

①

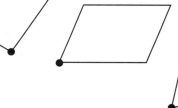

- **Square** •
- **Rectangle** •
- **Parallelogram** •
- **Trapezoid** •
- **Rhombus** •

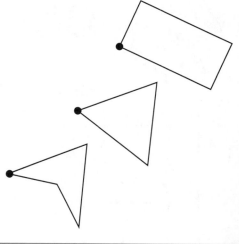

Look at the shapes the beaver made. Name the shapes. Then answer the questions.

②

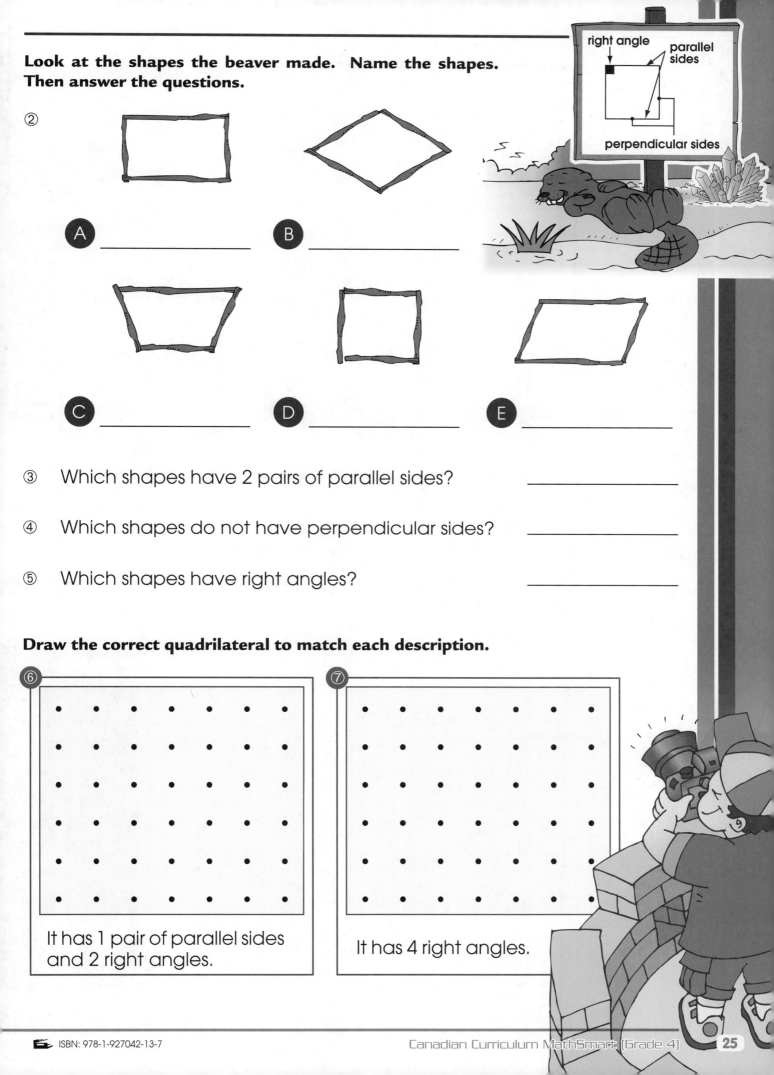

right angle | parallel sides

perpendicular sides

A _____

B _____

C _____

D _____

E _____

③ Which shapes have 2 pairs of parallel sides? _____

④ Which shapes do not have perpendicular sides? _____

⑤ Which shapes have right angles? _____

Draw the correct quadrilateral to match each description.

⑥

It has 1 pair of parallel sides and 2 right angles.

⑦

It has 4 right angles.

Write "congruent" or "similar" to describe each pair of shapes.

⑧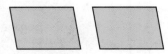

_____

⑨

_____

⑩

_____

⑪

_____

⑫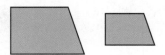

_____

⑬

_____

Draw a similar figure and a congruent figure for each shape.

⑭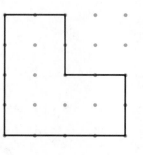

⑮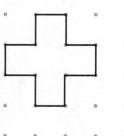

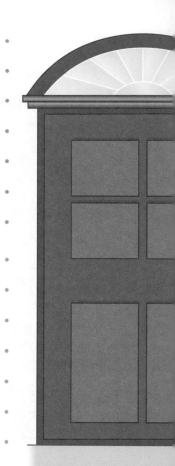

See how Tony measures the angles with a protractor. Then help him record the coloured angles in degrees.

vertex of an angle

arms of the angle

Degree (°) is a unit for measuring angles.

∠ AOB = 60°

To measure the angle, you should put the centre of the protractor at the vertex.

16

17

18

19

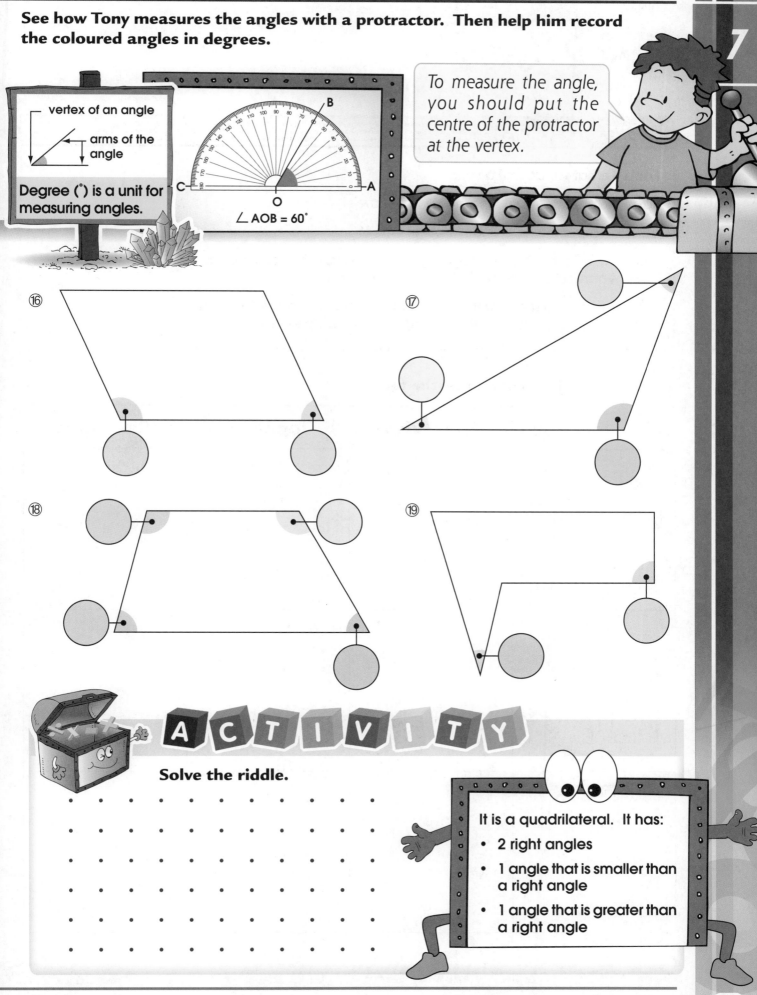

ACTIVITY

Solve the riddle.

It is a quadrilateral. It has:
- 2 right angles
- 1 angle that is smaller than a right angle
- 1 angle that is greater than a right angle

ISBN: 978-1-927042-13-7

8 3-D Figures

Face – a flat surface of a solid

Edge – a line segment where two faces meet

Vertex – a point where lines or edges meet

Prism – a solid shape with rectangular faces and congruent ends, named according to the shape of its end

Pyramid – a solid shape with triangular faces, a common vertex, and a flat base, named according to the shape of its base

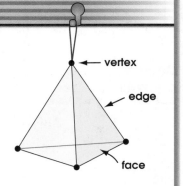

Name each solid. Then draw the faces.

	Front View	Top View	Side View
①			
②			
③			

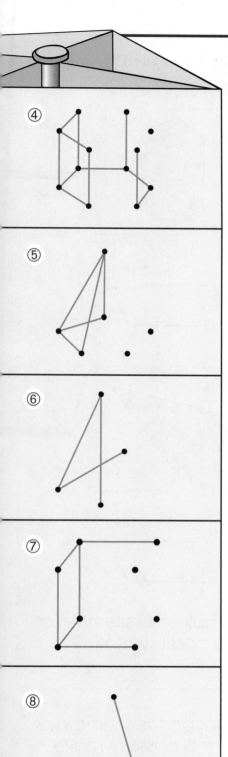

Draw lines to complete the skeleton of each 3-D figure. Then fill in the blanks.

④ A _____ has _____ edges, _____ vertices, and _____ faces.

⑤ A _____ has _____ edges, _____ vertices, and _____ faces.

⑥ A _____ has _____ edges, _____ vertices, and _____ faces. All the faces are in the shape of a _____ .

⑦ A _____ has _____ edges, _____ vertices, and _____ faces. All the faces are in the shape of a _____ .

⑧ A _____ has _____ edges, _____ vertices, and _____ faces. It has _____ triangular faces.

⑨ Each face has the same size and shape. They are in the shape of a _____ . It is a _____ .

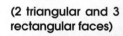

Triangular prism
- 9 edges
- 6 vertices
- 5 faces
(2 triangular and 3 rectangular faces)

ISBN: 978-1-927042-13-7

Write the numbers to tell how many marshmallows and sticks are needed to make the skeleton of each solid. Then answer the questions.

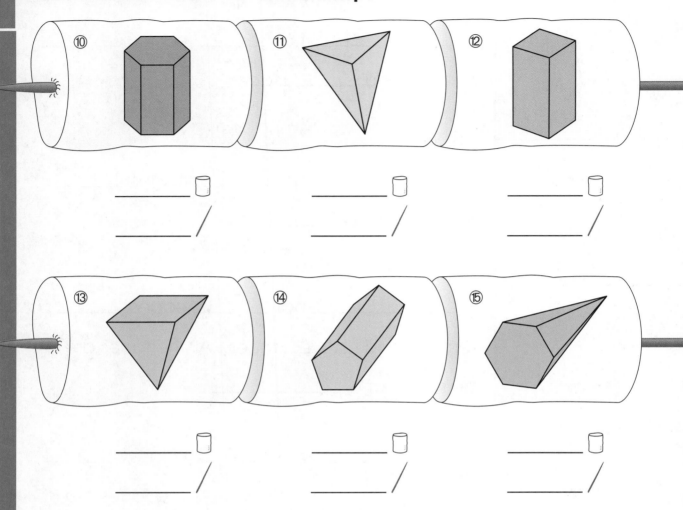

⑯ Jimmy uses 8 marshmallows and some sticks to make the skeleton of a 3-D figure shown above. Which figure can he possibly make?

_____

⑰ Diana uses 12 sticks and some marshmallows to make the skeleton of a 3-D figure shown above. Which 2 figures can she possibly make?

_____

⑱ How many marshmallows and sticks are needed to build a pentagonal pyramid?

⑲ How many marshmallows and sticks are needed to build a prism with a base having 7 edges?

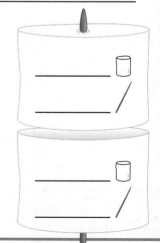

Canadian Curriculum MathSmart (Grade 4)   ISBN: 978-1-927042-13-7

Follow the methods to draw the pyramids or prisms.

Hexagonal pyramid

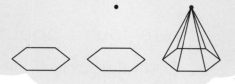

Triangular prism

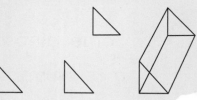

⑳ Rectangular prism

㉑ Pentagonal pyramid

Steps to draw a pyramid:
- Draw the base.
- Draw a point away from the base.
- Join each vertex of the base to the point.

㉒ Triangular pyramid

㉓ Hexagonal prism

Steps to draw a prism:
- Draw the end.
- Slide a congruent shape to the right and up a bit.
- Join the matching vertices.

A C T I V I T Y

Look at the table. Use the information to answer the questions.

① A pyramid has 18 sides in its base. It has

_____ edges and _____ vertices.

② A prism has 10 sides in its end. It has

_____ edges and _____ vertices.

	Pyramid		Prism	
No. of Edges	No. of sides in the base	x 2	No. of sides in the end	x 3
No. of Vertices	No. of sides in the base	+1	No. of sides in the end	x 2

Midway Test

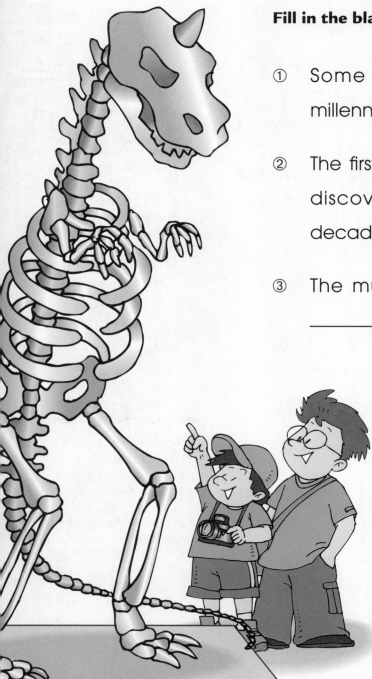

Fill in the blanks to complete the sentences. (5 marks)

① Some dinosaur fossils were found about 2 millenniums, or _____ centuries, ago.

② The first nearly-complete dinosaur skeleton was discovered about 2 centuries, or _____ decades, ago.

③ The museum was built about 8 decades, or _____ years, ago.

④ It is expected that the number of visitors to the museum will double in the next 20 years, or the next _____ decades.

⑤ Uncle John has worked at the museum for over 4 decades, or _____ years.

Look at the number of visitors to the museum in the last 3 months. Write the numbers in words. (6 marks)

No. of Visitors
May : 5410
June : 7899
July : 9301

⑥ May _____

⑦ June _____

⑧ July _____

Measure the angles. Then put them in order. (8 marks)

⑨

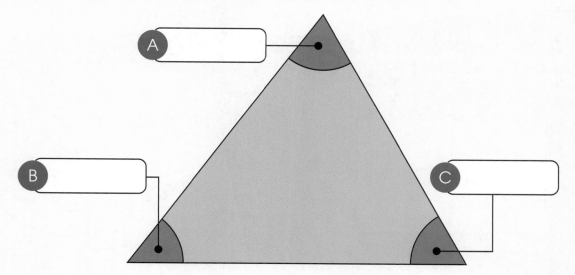

A

B

C

⑩　From smallest to biggest: _____

Find the answers. (14 marks)

⑪
$$
\begin{array}{r}
1717 \\
+\ \ \ 512 \\
\hline
\end{array}
$$

⑫
$$
\begin{array}{r}
1329 \\
+\ 8033 \\
\hline
\end{array}
$$

⑬
$$
\begin{array}{r}
2448 \\
-\ \ \ 327 \\
\hline
\end{array}
$$

⑭　$2001 - 226$ = _____　　⑮　$5254 + 868$ = _____

⑯　$2718 + 3899$ = _____　　⑰　$9011 - 917$ = _____

**Use addition to check the answers of the subtraction sentences.
Put a check mark ✔ in the box if it is correct. Write the correct
answer if it is not. (4 marks)**

⑱　$6313 - 167$ = __6146__　☐

　Check _____

⑲　$5000 - 555$ = __4555__　☐

　Check _____

Midway Test

Name each shape. Then measure and find its perimeter. (4 marks)

⑳

㉑

Perimeter: _____

Perimeter: _____

Find the congruent and similar pairs. Write the letters on the lines. (6 marks)

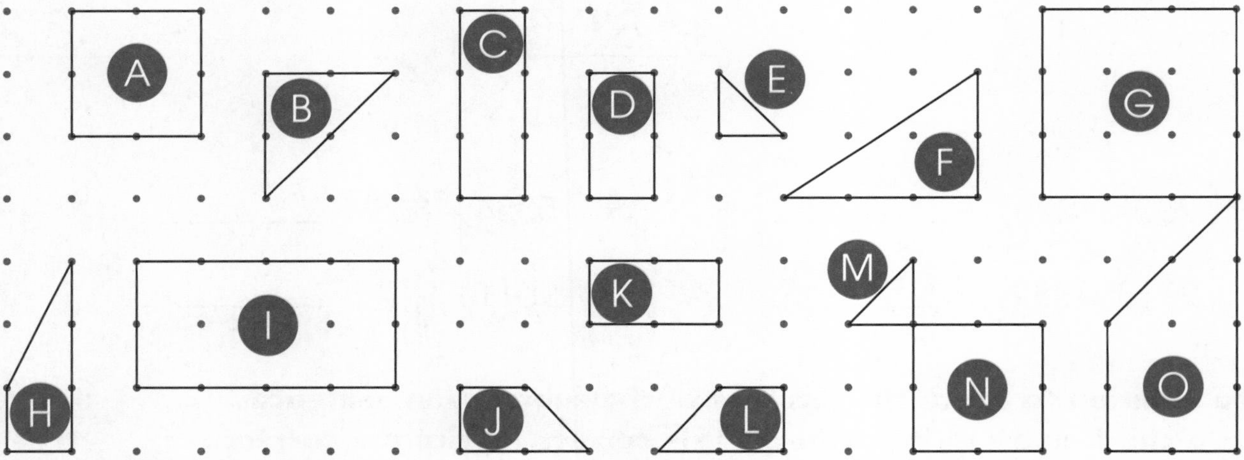

A B C D E F G

H I J K L M N O

㉒ **Congruent pairs**

A and _____

_____

_____

_____

Similar pairs

_____ _____

_____ _____

_____ _____

_____ _____

The gorilla looks at each 3-D figure from 3 different views. Help him name each figure. (6 marks)

Top View   **Front View**   **Side View**

㉓   _____

㉔   _____

㉕   _____

Fill in the missing numbers. (5 marks)

㉖ $4621 = 4000 + \rule{2cm}{0.4pt} + 20 + \rule{2cm}{0.4pt}$

㉗ $8989 = \rule{2cm}{0.4pt} + 900 + 80 + \rule{2cm}{0.4pt}$

㉘ $\rule{2cm}{0.4pt} = 5000 + 500 + 10 + 5$

Do the multiplication. (10 marks)

㉙
```
    8 5
  x   7
  _____
```

㉚
```
    9 2
  x   5
  _____
```

㉛
```
  1 4 6
  x   4
  _____
```

㉜
```
  2 9 9
  x   3
  _____
```

㉝
```
  1 2 5
  x   6
  _____
```

ISBN: 978-1-927042-13-7

Midway Test

Look at the picture. Name the shapes of the top and bottom parts of the stand. Then answer the questions. (8 marks)

top part

bottom part

㉞ Top part: _____

Bottom part: _____

㉟ 3:36   3:45

How long does it take Susan to make the skeleton?

_____ min

㊱ How many marshmallows and sticks are needed to build the skeleton of a rectangular pyramid?

_____ marshmallows and _____ sticks

㊲ Kevin traced the base of his 3-D figure on the grid. Help him find the area of the base.

1 cm

1 cm

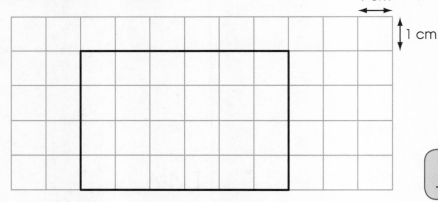

_____ cm²

Do the division. (12 marks)

㊳ 96 ÷ 4 = _____

㊴ 67 ÷ 5 = _____

㊵ 274 ÷ 2 = _____

㊶ 814 ÷ 7 = _____

㊷ 92 ÷ 6 = _____

㊸ 701 ÷ 3 = _____

Solve the problems. (12 marks)

	No. of Spectators
Yesterday	1728
Today	1697

44. How many spectators in total have there been for the 2 days?

_____ spectators in total

45. If 1198 of the spectators today are adults, how many children are there?

_____ children

46. There were 5 competitions last week. The same number of tickets was sold for each competition. If 9840 tickets were sold last week, how many tickets were sold for each competition?

_____ tickets

47. Each ticket costs $57. If Morris buys 4 tickets, how much does he need to pay?

$ _____

Score

100

9 Fractions

Fraction – a number showing a part of a whole

e.g. $\frac{1}{4}$ of the shape is coloured.

$\frac{1}{4}$ ← numerator ← denominator

Proper fraction – a fraction with a numerator smaller than its denominator

e.g. $\frac{3}{5}$ (3 < 5)

Improper fraction – a fraction with a numerator greater than its denominator

e.g. $\frac{4}{1}$ (4 > 1)

Mixed number – a number formed by a whole number and a proper fraction

e.g. $2\frac{4}{7}$ (2 and $\frac{4}{7}$)

Look at the stars. Answer the questions.

① What fraction of the stars are

    a. yellow? _____

    b. green? _____

    c. blue? _____

Sort the fractions.

② Proper fraction

_____

③ Improper fraction

_____

④ Mixed number

_____

Balloons: $\frac{4}{7}$ $3\frac{1}{6}$ $\frac{9}{2}$ $\frac{11}{3}$ $\frac{5}{14}$ $\frac{1}{5}$ $1\frac{1}{9}$ $\frac{3}{8}$ $\frac{2}{1}$ $2\frac{5}{7}$ $\frac{13}{6}$ $1\frac{4}{5}$

Write the mixed number represented by each group of diagrams.

⑤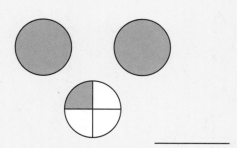

_____

⑥

_____

⑦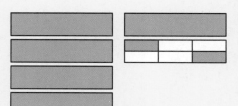

_____

⑧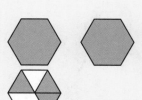

_____

2 wholes   3 parts coloured

Mixed number:
$2\frac{3}{8}$ coloured

Look at the flowers. Fill in the blanks with fractions.

⑨ _____ of the flowers are yellow.

⑩ _____ of the flowers are green.

⑪ _____ of the flowers are red.

Colour the diagrams to show the fractions.

⑫ 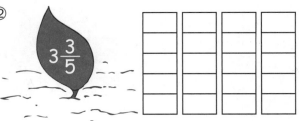 $3\frac{3}{5}$

⑬ $\frac{7}{3}$

For fractions with the same denominator, compare their numerators.

e.g. $\dfrac{4}{5} > \dfrac{2}{5}$ ← 4 > 2 ← same

Mixed numbers with the same denominator
Compare the whole number parts first. If they are the same, compare their numerators.

Compare each pair of fractions. Put ">" or "<" in the circle.

⑭ $\dfrac{4}{7} \bigcirc \dfrac{6}{7}$

⑮ $\dfrac{9}{4} \bigcirc \dfrac{3}{4}$

⑯ $1\dfrac{5}{8} \bigcirc 4\dfrac{7}{8}$

⑰ $\dfrac{4}{9} \bigcirc \dfrac{1}{9}$

⑱ $\dfrac{7}{5} \bigcirc \dfrac{10}{5}$

⑲ $8\dfrac{2}{5} \bigcirc 8\dfrac{3}{5}$

⑳ $1\dfrac{3}{4} \bigcirc 3\dfrac{1}{4}$

㉑ $\dfrac{2}{3} \bigcirc \dfrac{1}{3}$

㉒ $2\dfrac{3}{7} \bigcirc 3\dfrac{2}{7}$

㉓ $1\dfrac{7}{10} \bigcirc 1\dfrac{9}{10}$

㉔ $2\dfrac{4}{5} \bigcirc 1\dfrac{4}{5}$

㉕ $\dfrac{5}{8} \bigcirc 1\dfrac{1}{8}$

Colour the diagram to show each fraction. Then put the fractions in order.

㉖ $\dfrac{5}{6}$ $\dfrac{1}{6}$ $\dfrac{3}{6}$

From greatest to smallest: _____

㉗ $1\dfrac{1}{3}$ $1\dfrac{2}{3}$ $2\dfrac{1}{3}$

From smallest to greatest: _____

Put the fractions in order from smallest to greatest.

㉘ $\dfrac{3}{7}$ $\dfrac{6}{7}$ $\dfrac{2}{7}$

_____

㉙ $5\dfrac{1}{6}$ $6\dfrac{5}{6}$ $5\dfrac{3}{6}$

_____

Draw and colour pictures to show the things in each group. Then fill in the blanks with fractions.

㉚ Ricky can see 12 🌸 on the mountain. $\frac{2}{12}$ of the flowers are red. $\frac{7}{12}$ are yellow. The rest are purple.

_____ of the flowers are purple.

㉛ Ricky can see 10 🪁. $\frac{5}{10}$ of the kites are red. $\frac{2}{10}$ are blue. The rest are green.

_____ of the kites are green.

ACTIVITY

There are 20 children in Mrs. Lynn's class. Help Mrs. Lynn draw the faces and hats. Then answer the question.

- $\frac{3}{5}$ of the children have a happy face 🙂 ; the rest have a sad face 🙁 .

- $\frac{1}{2}$ of the children wear a hat.

At least how many children have a happy face and wear a hat?

_____ children

10 Decimals

WORDS TO LEARN

Decimal – a numeral containing a decimal point, with the value of the digit(s) to the right of the decimal point being less than 1

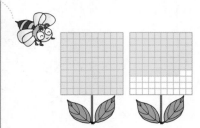

Ones   Tenths   Hundredths

$$1.68$$

decimal point

In words: One and sixty-eight hundredths

In fraction: $1 \frac{68}{100}$

Write a fraction and a decimal for each diagram to show how much is coloured.

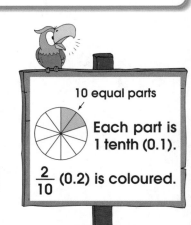

10 equal parts

Each part is 1 tenth (0.1).

$\frac{2}{10}$ (0.2) is coloured.

①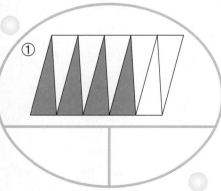

②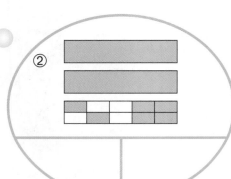

③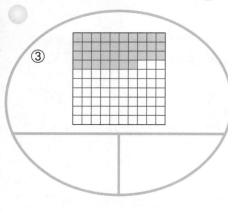

④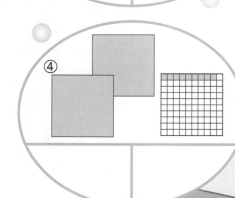

Write the equivalent decimals or fractions.

⑤ $\frac{3}{10}$ = _____

⑥ $\frac{51}{100}$ = _____

⑦ $\frac{2}{100}$ = _____

⑧ $1\frac{9}{10}$ = _____

⑨ 1.69 = _____

⑩ 0.8 = _____

⑪ 5.03 = _____

⑫ 4.6 = _____

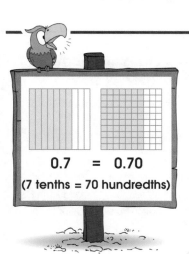

0.7 = 0.70
(7 tenths = 70 hundredths)

Write as a decimal.

⑬ Five tenths _____

⑭ Twenty hundredths _____

⑮ Three hundredths _____

⑯ One and seven tenths _____

⑰ Sixty-one hundredths _____

⑱ Two and twelve hundredths _____

Write the decimals in words.

⑲ 0.9 _____

⑳ 0.58 _____

㉑ 1.04 _____

㉒ 7.23 _____

Write the place value and meaning of each coloured digit.

㉓ 3.72 _____ ; _____

㉔ 0.69 _____ ; _____

㉕ 3.54 _____ ; _____

㉖ 8.41 _____ ; _____

9.58
5 is in the tenths column.
It means 0.5.
_____
4.5 = 4.50
They have the same value.

Circle the pair of decimals with the same value in each group.

㉗ 3.004
3.04   3.40
3.4

㉘ 0.205
0.25   0.250
0.025

㉙ 0.6
0.60   0.06
0.006

㉚ 0.08
0.8   8.00
0.080

ISBN: 978-1-927042-13-7

Circle the item with the lower price in each group. Then use the same colour to colour the matching item on Lucy.

㉛ $29.4 $24.9

㉜ $2.39 $2.59

㉝ $8.15 $7.99

㉞ $0.5 $0.65

㉟ $17.95 $17.99

You may compare the decimals using a number line.
e.g.

1.6   1.64   1.7 1.72

1.64 < 1.72

Look at the decimals on the T-shirt. Find and write the decimals on the lines.

㊱ smaller than 4.15: _____

㊲ between 3.8 and 5.5: _____

㊳ has an "8" in its tenths column:

_____

㊴ Put the decimals in order. Start with the smallest.

_____

_____

7.15   6.6

2.86   8.14

3.4   3.84

4.7   3.98

Write the children's heights in metres. Then answer the questions.

㊵ Jimmy   1 m 9 cm   = _____ m

㊶ Aaron   1 m 12 cm = _____ m

㊷ Fiona   98 cm   = _____ m

㊸ Jerry   1 m 3 cm   = _____ m

㊹ Nancy is 2 cm taller than Jerry. She is _____ m.

㊺ Linda is 4 cm shorter than Aaron. She is _____ m.

㊻ Who is the tallest?   _____

㊼ Who is the shortest?   _____

㊽ Put the children in order. Start with the shortest.

_____ , _____ , _____ ,

_____ , _____ , _____

No cheating when you are measuring your height!

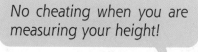

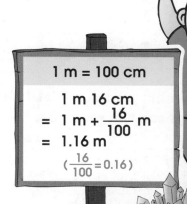

1 m = 100 cm

$1 \text{ m } 16 \text{ cm}$
$= 1 \text{ m} + \dfrac{16}{100} \text{ m}$
$= 1.16 \text{ m}$
$\left(\dfrac{16}{100} = 0.16 \right)$

A C T I V I T Y

Look at Jimmy's picture. Fill in the blanks with decimals to complete the sentences.

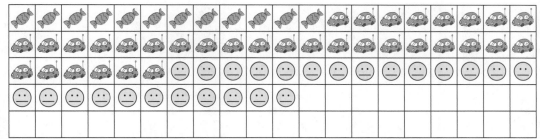

① _____ of the picture is 🍬 .

② _____ of the picture is 🚗 .

③ _____ of the picture is 😐 .

④ _____ of the picture is blank.

11 Addition and Subtraction with Decimals

You can estimate the sums or differences by rounding each number to the nearest whole number. Then add or subtract the rounded numbers.

3.1   4.7

2   3   4   5   6

4.7 is rounded to 5.
3.1 is rounded to 3.

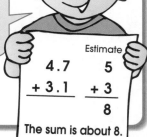

Estimate

$$\begin{array}{r} 4.7 \\ + 3.1 \\ \hline \end{array} \qquad \begin{array}{r} 5 \\ + 3 \\ \hline 8 \end{array}$$

The sum is about 8.

Write the decimals and complete the addition and subtraction sentences to match the pictures.

①

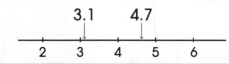

_____ + _____ = _____

②

_____ - _____ = _____

③

_____ + _____ = _____

④

_____ - _____ = _____

⑤

_____ + _____ = _____

Fill in the correct digits.

⑥ **Addition**

1st Align the decimal points.	2nd Add the hundredths.	3rd Add the tenths.	4th Add the ones.
1 . 3 2 + 0 . 5 9	1 . 3 2 + 0 . 5 9 ☐	1 . 3 2 + 0 . 5 9 ☐ 1	1 . 3 2 + 0 . 5 9 ☐ . 9 1

⑦ **Subtraction**

1st Align the decimal points.	2nd Subtract the hundredths.	3rd Subtract the tenths.	4th Subtract the ones.
1 . 5 6 − 0 . 4 8	1 . 5 6 − 0 . 4 8 ☐	1 . 5 6 − 0 . 4 8 ☐ 8	1 . 5 6 − 0 . 4 8 ☐ . 0 8

Find the answers.

⑧
$$0.68 + 0.22$$

⑨
$$7.14 + 2.89$$

⑩
$$6.6 + 2.48$$

⑪
$$9.14 - 0.09$$

⑫
$$2.01 - 0.13$$

⑬
$$3.59 - 2.7$$

⑭ 3.03 + 1.59 = _____

⑮ 6.1 + 4.95 = _____

⑯ 9 − 1.08 = _____

⑰ 3.16 + 8.07 = _____

⑱ 5.13 + 2.87 = _____

⑲ 7.6 − 5.38 = _____

⑳ 1.16 + 4.74 = _____

㉑ 8.15 − 7.77 = _____

Add or subtract as usual. Remember to put the decimal point in the answer directly under the decimal points above.

ISBN: 978-1-927042-13-7   Canadian Curriculum MathSmart (Grade 4)

Round each decimal to the nearest whole number to do the estimate. Then find the answer.

㉒ 2.43 + 1.57 = _____   (Estimate) _____

㉓ 5.71 – 0.94 = _____   (Estimate) _____

㉔ 8.26 + 1.3 = _____

(Estimate) _____

㉕ 9.1 – 1.49 = _____

(Estimate) _____

㉖ 0.52 + 1.89 = _____

(Estimate) _____

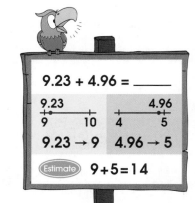

9.23 + 4.96 = _____

9.23 → 9   4.96 → 5

(Estimate) 9 + 5 = 14

Look at the map. Help Mr. Brown solve the problems.

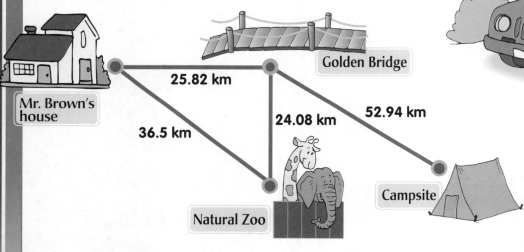

Mr. Brown's house — 25.82 km — Golden Bridge

36.5 km

24.08 km

52.94 km

Natural Zoo

Campsite

For questions ㉗ and ㉘, use addition to solve the problems.

For question ㉙, use subtraction.

㉗ If Mr. Brown drives from his house to the campsite crossing the Golden Bridge, he will drive _____ km in all.

㉘ If a tiger escapes from the zoo to the Golden Bridge and then returns, it will have run _____ km in total.

㉙ Mr. Brown and his family are on their way to the Natural Zoo. If Mr. Brown has driven 27.56 km already, he will need to drive _____ km more.

Find the total points each dog gets at the Annual Dog Show. Then answer the questions.

㉚
 1st round  2nd round   1st round 7.68  2nd round 8.94   1st round  2nd round

8.85   9.23   8.5   9.72

Pal: _____ points   Cookie: _____ points   Beauty: _____ points

㉛ Which dog is the champion? _____

㉜ The top score for each round is 10 points. How many points did Pal lose in the two rounds?

_____ = _____

_____ points

㉝ Cookie got 9.24 points in the 1st round but he was penalized for cheating. How many points were deducted?

_____ = _____

_____ points

Dog show

 A C T I V I T Y

Each child, together with his or her pet, weighs 44.81 kg. Match the children with their pets. Write the letters.

35.21 kg   34.02 kg   34.07 kg

① ② ③

Ⓐ 9.6 kg

Ⓑ  10.74 kg

Ⓒ  10.79 kg

ISBN: 978-1-927042-13-7   Canadian Curriculum MathSmart (Grade 4)

12 Money

WORDS TO LEARN

$ – dollar sign

¢ – cent sign

$2.78 means 2 dollars 78 cents.

Dollars Cents

$2.78 =

Help the children find out how much each cake costs. Estimate first. Then write the amount in dollars.

①

Estimate: _____

Actual
Amount: $ _____

②

Estimate: _____

Actual
Amount: $ _____

Many items at ABC Toy Shop are on sale now. Write how much can be saved and the reduced prices.

③

was $16.32

save $ _____

Now $ _____

④

was $36.50

save $ _____

Now $ _____

⑤

was $9.82

save $ _____

Now $ _____

⑥

was $50.84

save $ _____

Now $ _____

SALE

$5 – $15	$5 off
$16 – $30	$10 off
$31 – $45	$15 off
$46 – $60	$20 off

Help Mr. Cowie, the pet shop owner, write the sale price of each animal on the board. Then answer the questions.

Birds $47.65 each

⑦ SALE PRICE

🐟 ┈┈┈┈ $

🐠 ┈┈┈┈ $

🐢 ┈┈┈┈ $

Birds $

each

under $20 $2.25 off
over $20 $4.88 off

$12.37

$17.70

$15.72

⑧ If Mrs. Lynn buys a 🐟 and a 🐠 , how much does she need to pay in all?

_____ = _____ $_____

⑨ Julie has $50. If Julie buys a bird, how much will be left?

_____ = _____ $_____

⑩ Jimmy and David like turtles. If they each buy a turtle, how much do they need to pay in all?

_____ = _____ $_____

ACTIVITY

This is Josephine's money. How much more does she need to equal $50? What are the fewest bills and coins needed to get this amount?

ISBN: 978-1-927042-13-7

13 Capacity, Volume, and Mass

WORDS TO LEARN

Capacity – the amount of water a container can hold
There are 1000 millilitres (mL) in a litre (L); 1 L = 1000 mL.

Volume – the amount of space an object takes up
The volume of a centimetre cube is 1 cubic centimetre (cm^3).

Mass – Kilogram (kg), gram (g), and milligram (mg) are the units for measuring mass.

1 kg = 1000 g; 1 g = 1000 mg

Volume = 1 cm^3

Choose the more appropriate unit to measure the capacity of each container. Write "L" or "mL".

① A bucket _____

② A cup _____

③ A bathtub _____

④ A can of soup _____

⑤ A washing machine _____

⑥ A cargo ship _____

L – a big unit
mL – a small unit

1 L = 1000 mL

4 L = (4 x 1000) mL
= 4000 mL

Fill in the blanks.

⑦ 5 L = _____ mL

⑧ 7000 mL = _____ L

⑨ 9000 mL = _____ L

⑩ 3 L = _____ mL

⑪ 1 L 20 mL = _____ mL

⑫ 5 L 400 mL = _____ mL

⑬ 2 L 8 mL = _____ mL

⑭ 3 L 86 mL = _____ mL

Write the capacity of each container on the line.

Capacities

⑮

_____ L

_____ mL

_____ mL

_____ mL

Circle the correct answers.

⑯ The capacity of a container is 1200 mL. (2   3   4) cans of juice are needed to fill it up.

⑰ The bucket is fully filled with water. If half of the water in the bucket is poured equally into 4 bottles, each bottle will contain (500 mL   1 L   250 mL) of water.

⑱ If Aunt Lucy adds 3 cans of soup and 460 mL of water into the pot, there will be (1 L   820 mL   80 mL) of liquid in the pot.

13

The structures are built with centimetre cubes. Find the volume of each structure.

⑲

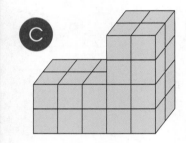 A

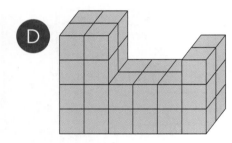 B

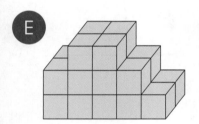 C

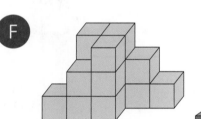 D

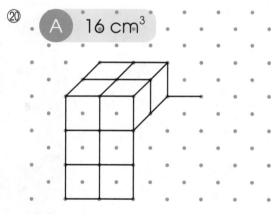 E

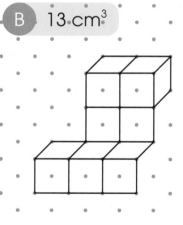 F

A _____ cm³

B _____ cm³

C _____ cm³

D _____ cm³

E _____ cm³

F _____ cm³

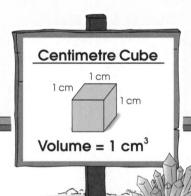

Centimetre Cube

1 cm × 1 cm × 1 cm

Volume = 1 cm³

Complete the diagrams to match the specified volumes. Then answer the questions.

⑳ A **16 cm³**   B **13 cm³**

㉑ How many cubes cannot be seen in A ? _____

㉒ How many cubes cannot be seen in B ? _____

Help each dog find its weight. Then answer the questions.

㉓ A kg   B kg   C kg

1 kg = 1000 g

$\frac{1}{2}$ kg = 500 g

1 kg 8 g = 1000 g + 8 g
= 1008 g

㉔ Dog _____ is the heaviest.

㉕ Dog _____ is the lightest.

㉖ Dog C is _____ kg heavier than Dog B.

㉗ If Dog A weighs 9130 g with a bone in its mouth, the bone weighs _____ g.

㉘ If Dog A has gained $\frac{1}{2}$ kg in the past two months, it weighed _____ g two months ago.

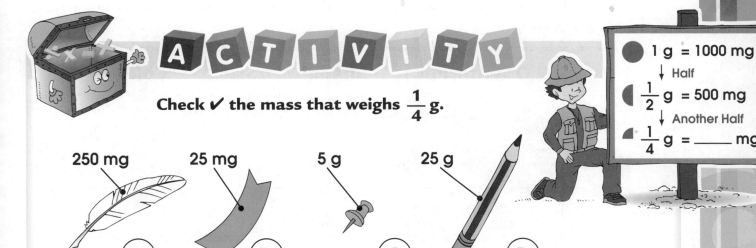

ACTIVITY

Check ✔ the mass that weighs $\frac{1}{4}$ g.

1 g = 1000 mg
↓ Half
$\frac{1}{2}$ g = 500 mg
↓ Another Half
$\frac{1}{4}$ g = _____ mg

250 mg — (A)   25 mg — (B)   5 g — (C)   25 g — (D)

ISBN: 978-1-927042-13-7

14 Patterning

Help Susan draw the next 2 pictures in each pattern.

①

②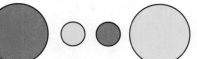

③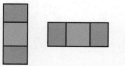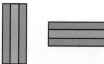

Choose the given words to describe which two attributes change in each pattern.

Orientation	Shape	Size	Colour

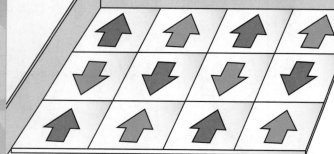 

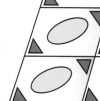

④ _____ ; _____ ⑤ _____ ; _____

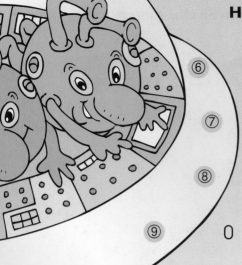

Help the aliens fill in the missing numbers.

⑥ 13 17 21 ☐ 29 33 37 ☐

⑦ 6 12 18 24 ☐ ☐ 42 48

⑧ 90 84 ☐ 72 ☐ 60 54 48

⑨ 0 15 30 ☐ 60 ☐ 90 105 ☐

⑩ 2 5 9 14 20 ☐ ☐ 44 ☐

Write the next 2 numbers in each pattern. Then write the rule for each pattern.

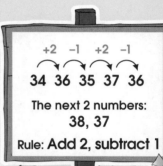

+2 −1 +2 −1
34 36 35 37 36
The next 2 numbers:
38, 37
Rule: Add 2, subtract 1

⑪ 50 42 52 44 54 _____ _____

Rule: _____

⑫ 8 16 13 26 23 _____ _____

Rule: _____

⑬ 4 16 8 32 16 _____ _____

Rule: _____

Follow the patterns to complete the number sentences.

⑭ 15 + 7 = 6 + 16

15 + 8 = 6 + 17

15 + 9 = 6 + ____

15 + ____ = 6 + ____

____ + ____ = 6 + ____

⑮ 29 − 4 = 20 + 5

29 − 5 = 20 + 4

29 − 6 = 20 + ____

29 − ____ = 20 + ____

____ − ____ = 20 + ____

ISBN: 978-1-927042-13-7

Follow the patterns to draw the 4th groups. Then answer the questions.

1st Group   **2nd Group**   **3rd Group**   **4th Group**

⑯

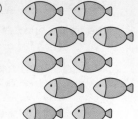  

⑰

⑱

⑲ a. The number of flowers (increases decreases) by _____ each time.

b. There are _____ flowers in the 5th group; _____ flowers are red and _____ are yellow.

⑳ a. The number of stones (increases decreases) by _____ each time.

b. There are _____ stones in the 5th group.

㉑ a. The number of fish (increases decreases) by _____ each time.

b. There are 2 fish in the _____ group.

You can use a number pattern to represent the pictures in a pattern. e.g.

: :: ::: (Goes up by
2 → 4 → 6   2 each time.)

Read what Michelle says. Then complete the table and answer the questions.

Each child is allowed to pick 6 red and 9 green apples.

㉒

No. of Children	1	2	3	4	5	6
Total No. of 🍎	6					
Total No. of 🍏	9					
Total No. of Apples	15					

㉓ The number of red apples picked increases by _____ as the number of children goes up by 1 each time.

㉔ How many green apples will be picked by 7 children? _____

㉕ How many apples will be picked by 8 children in all? _____

㉖ Follow the pattern to draw and colour the apples in the 4th group.

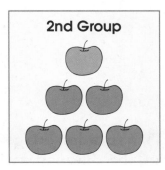
1st Group

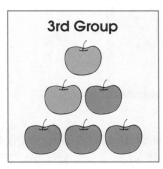
2nd Group

3rd Group

4th Group

ACTIVITY

Solve the problem.

There are 5 apples in the first basket. If the number of apples in each basket doubles that of the previous one, how many apples are there in the first 4 baskets?

_____ apples

Make a table to solve this problem.

Basket	1st	2nd
No. of 🍎	5	10

15 Transformation and Coordinates

WORDS TO LEARN

Transformation – a change in position or direction of a figure by translation (slide), rotation (turn), or reflection (flip)

Translation – sliding a figure up, down, or sideway

Rotation – turning a figure about a fixed point

Reflection – flipping a figure over a line

A figure does not change its size and shape under the transformations – translation, rotation, and reflection.

Translation

Rotation

$\frac{1}{2}$ turn

turning point

Reflection

♥ is at (3,4).

Coordinates – an ordered pair used to describe a location on a grid; the order of a pair (units across, units up)

Colour the translation images of the flower green, the rotation images red, and the reflection images yellow.

①

Draw a translation image (T), a rotation image (R), and a reflection image (E) for each shape. Then label the images with "T", "R", and "E".

②

③

Help the pirate answer the questions.

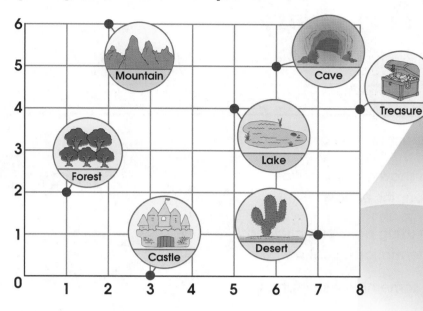

The coordinates of A are (2,3).

A(2,3)

3 units up

2 units across

④ What do you find at (3,0)? _____

⑤ What do you find at (6,5)? _____

⑥ Where is the mountain? _____

⑦ Where is the treasure? _____

⑧ The castle is _____ units up down and _____ units left right from the cave.

⑨ There is a bridge 4 units down and 2 units right from the mountain. Draw a ● to show the location of the bridge on the map.

ACTIVITY

Flip the robot over the red line. Then find the new coordinates of his eyes.

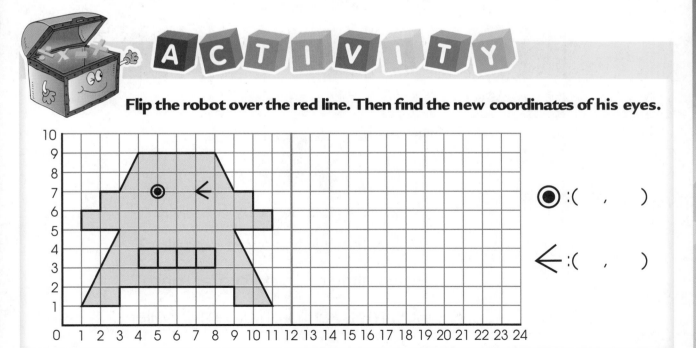

⦿ :(,)

← :(,)

16 Graphs and Probability

Tally – a method to record the numbers in groups of 5 卌

Bar graph — a graph showing information by the lengths of bars

Circle graph — a graph using parts of a circle to show information about a whole

Probability – the chance that something will happen

Tree diagram – a diagram that shows all the possible outcomes

My dog will speak tomorrow.

It is impossible!

See what the children in Mr. Miller's class like to have for breakfast. Read the graph. Then complete the table and answer the questions.

I love sausages!

The Children's Breakfast ☺ = 2 children

Bacon   Ham   Eggs   Sausages

☺ = 2 children

◗ = 1 child

①

Bacon	卌 ||
Ham	
Eggs	
Sausages	

② Which food is liked by 5 children? _____

③ Which food is the least popular? _____

④ How many children prefer meat? _____

⑤ How many children are in Mr. Miller's class? _____

Look at the results of the survey about the kind of movies the children like. Complete the bar graph to show the information and answer the questions.

⑥

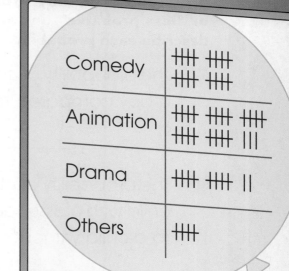

Comedy — HHT HHT HHT HHT
Animation — HHT HHT HHT HHT HHT III
Drama — HHT HHT II
Others — HHT

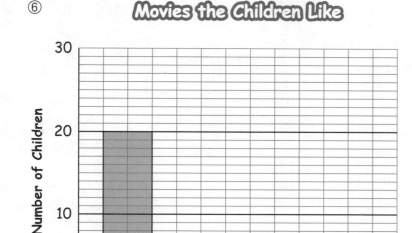

Movies the Children Like

(bar graph with vertical axis "Number of Children" 0 to 30, showing bar for "Comedy" at 20; horizontal axis "Kinds of Movies")

⑦ How many children like "Animation"? _____

⑧ How many fewer children like "Drama" than "Comedy"? _____

⑨ How many children were asked in all? _____

⑩ What is the label for the vertical axis? _____

⑪ What is the label for the horizontal axis? _____

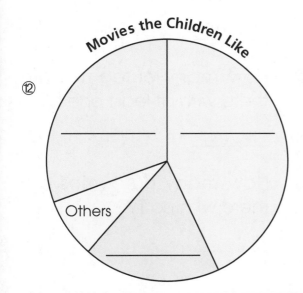

Movies the Children Like

⑫

(circle graph with sectors, one labelled "Others")

Help David use a circle graph to show the results of the survey above. Label each sector of the circle. Then answer the question.

⑬ What does "Others" refer to? Suggest an example that may be in this section.

_____

_____

16

Mrs. Kennedy wants to choose a student to be the project leader. Write "more probable", "equally probable", or "less probable" on the line to describe each probability.

⑭ The probability of choosing a boy compared to a girl:

_____

⑮ The probability of choosing a child with glasses compared to a child without glasses:

_____

⑯ The probability of choosing a child wearing a red T-shirt compared to a child not wearing a red T-shirt:

_____

The children are going to toss a coin twice and guess what they will get. List all the possible outcomes using a tree diagram. Then answer the questions.

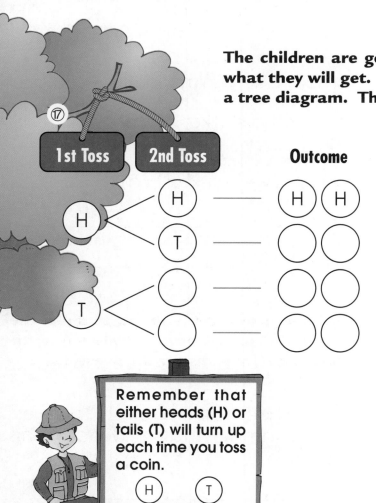

⑰

| 1st Toss | 2nd Toss | | Outcome |

⑱ How many outcomes are there in all?

_____

⑲ Is it more probable to get HH compared to TT?

_____

⑳ How many outcomes are there with at least one H?

_____

㉑ How many outcomes are there with no T?

_____

Remember that either heads (H) or tails (T) will turn up each time you toss a coin.

Wayne throws a dice and tosses a coin. Help him complete the tree diagram to show all the possible outcomes. Then answer the questions.

There are 6 faces on a dice. The number of dots on the faces are:

• (1) ⁝ (2) ∴ (3) ∷ (4) ⁙ (5) ∷∷ (6)

㉒
Dice	Coin	Outcome
1	H	1 H
	T	1 T
2	H	2 H
3		
4		
5		
6		

㉓ How many possible outcomes are there in all? _____

㉔ How many outcomes are with an H? _____

㉕ How many outcomes are with a number greater than 4? _____

㉖ Is it more probable to get 3H compared to 4T? _____

㉗ If Wayne repeats the game 120 times, about how many times will each outcome occur? Explain.

_____

_____

ACTIVITY

Read what Eric says. Answer the question.

If I do the same experiment as Wayne did, will the results be similar to Wayne's? Explain.

_____

Final Test

Colour the leaves as specified. (6 marks)

① $\frac{8}{9}$ $\frac{5}{4}$ $\frac{10}{11}$ $2\frac{1}{3}$ $\frac{9}{7}$ $4\frac{9}{10}$

Proper fraction – green
Improper fraction – red
Mixed number – blue

Write each fraction as a decimal or each decimal as a fraction. (6 marks)

② $\frac{7}{10}$ _____

③ 0.03 _____

④ 1.11 _____

⑤ $2\frac{29}{100}$ _____

⑥ 4.7 _____

⑦ $8\frac{9}{10}$ _____

Write the value of each coloured digit. (4 marks)

⑧ 16.78 _____

⑨ 22.33 _____

⑩ 10.56 _____

⑪ 54.99 _____

Follow the pattern to draw the next two pictures. Then circle the words to tell which two attributes change in the pattern. (4 marks)

⑫

Attribute change: Shape Size Colour Orientation

ISBN: 978-1-927042-13-7

Colour the greater value in each group. (3 marks)

⑬  $\dfrac{7}{8}$ $\dfrac{5}{8}$

⑭ 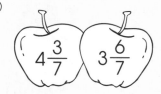 $4\dfrac{3}{7}$ $3\dfrac{6}{7}$

⑮  4.05 4.50

Find the answers. (5 marks)

⑯
$$\begin{array}{r} 5.16 \\ -\ 2.07 \\ \hline \end{array}$$

⑰
$$\begin{array}{r} 12.37 \\ +\ 6.66 \\ \hline \end{array}$$

⑱
$$\begin{array}{r} 9 \\ -\ 1.55 \\ \hline \end{array}$$

⑲ $3.7 + 2.88 =$ _____

⑳ $2.2 - 0.87 =$ _____

Look at the drinks. Then answer the questions. (4 marks)

 Milk $2.69

 Juice 255 ml $0.95

 Water $4.15

㉑ Two cans of juice are needed to fill up a carton. Is the capacity of the carton more or less than 1 L? _____

㉒ If May buys 1 carton of milk and 1 bottle of water with a $20 bill, what will be the change? $ _____

Final Test

Help Lily find how many more centimetre cubes are needed to fill up each container. Then find the volume of each container. (8 marks)

㉓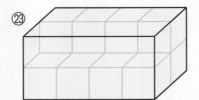

a. _____ more cubes are needed.

b. The volume of the container is _____ centimetre cubes.

㉔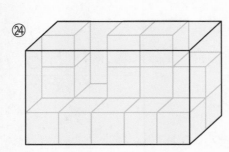

a. _____ more cubes are needed.

b. The volume of the container is _____ centimetre cubes.

Use a bar graph to record the number of each kind of solid on the shelves. Then answer the questions. (5 marks)

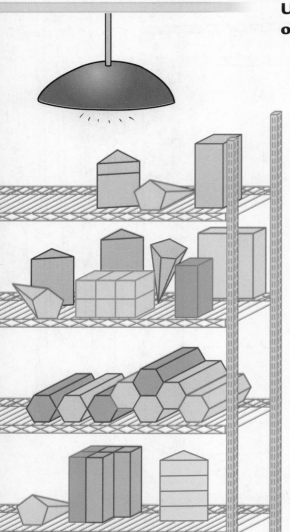

㉕

<u>Solids on the Shelves</u>

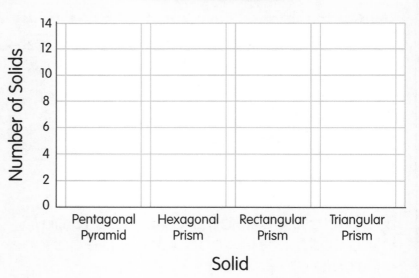

Number of Solids

Pentagonal Pyramid   Hexagonal Prism   Rectangular Prism   Triangular Prism

Solid

㉖ How many solids are there with 5 faces?

_____

㉗ How many solids are there with 12 edges?

_____

Look at the position of each child's house on the map. Help the children solve the problems. (10 marks)

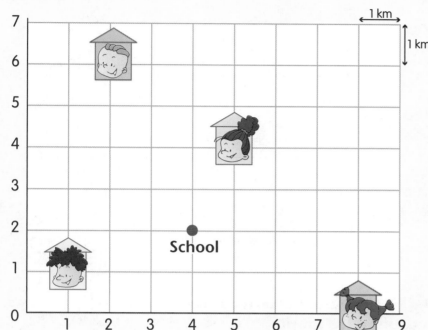

My name is Jane.

㉘ The coordinates of the children's houses:

a. _____   b. _____

c. _____   d. _____

㉙ Each child takes the shortest route to school. If each child is allowed to go vertically and horizontally, draw lines on the map to show their routes.

㉚ What is the travelling distance from Jane's house to her school? _____ km

㉛ If a supermarket is 3 units up and 5 units left from Jane's house, what are the coordinates of the supermarket? _____

Find the answers. (8 marks)

㉜ 248 x 4 = _____   ㉝ 717 ÷ 6 = _____

㉞ 477 x 5 = _____   ㉟ 815 ÷ 3 = _____

Final Test

Nicky the Dog is going to pick an apple. Help him write "more probable", "equally probable", or "less probable" to describe each situation. (6 marks)

36. The probability of picking a red apple compared to a green one: _____

37. The probability of picking a golden apple compared to a red one: _____

38. The probability of picking a green apple compared to a golden one: _____

Each bag of apples weighs 1 kg. Follow the pattern to complete the table. Then answer the questions. (8 marks)

39.

No. of Bags	No. of 🍎
1	7
2	14
3	21
4	28
5	
6	
7	
8	
9	

40. How many apples are there in 8 bags?

    _____ apples

41. If Mrs. Fred buys 70 apples, how many bags of apples does Mrs. Fred buy in all?

    _____ bags

42. How many bags of apples weigh 4000 g?

    _____ bags

Describe each transformation using "translation", "reflection", or "rotation". (3 marks)

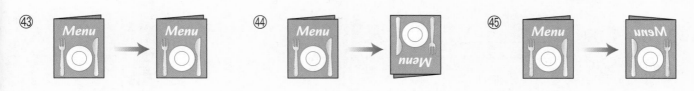

43 _____ 44 _____ 45 _____

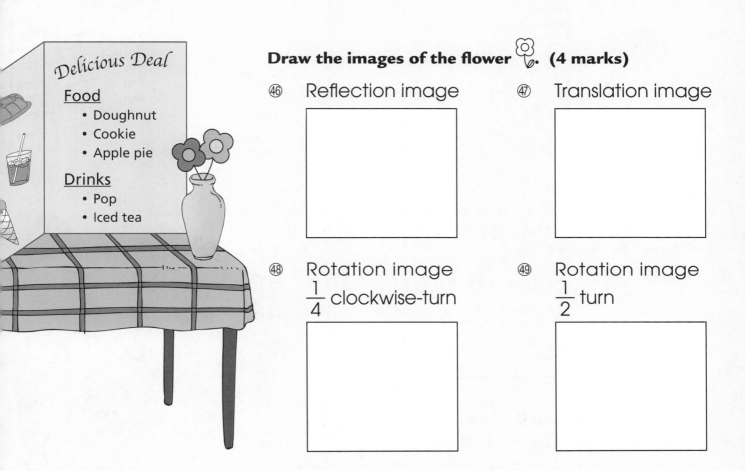

Draw the images of the flower. (4 marks)

46 Reflection image

47 Translation image

48 Rotation image
$\frac{1}{4}$ clockwise-turn

49 Rotation image
$\frac{1}{2}$ turn

Delicious Deal

Food
• Doughnut
• Cookie
• Apple pie

Drinks
• Pop
• Iced tea

Look at the menu above. Choose one item from each category. Show all the possible outcomes using a tree diagram. (4 marks)

50

Doughnut	(D)
Cookie	(C)
Apple pie	(A)
Pop	(P)
Iced tea	(I)

Food   Drink   Outcome

D

C

A

Final Test

Estimate by rounding each decimal to the nearest whole number. Then find the exact answer. (6 marks)

(51) 3.78 + 1.09 = ________

Estimate ________

(52) 0.87 + 4.21 = ________

Estimate ________

(53) 8.8 + 7.78 = ________

Estimate ________

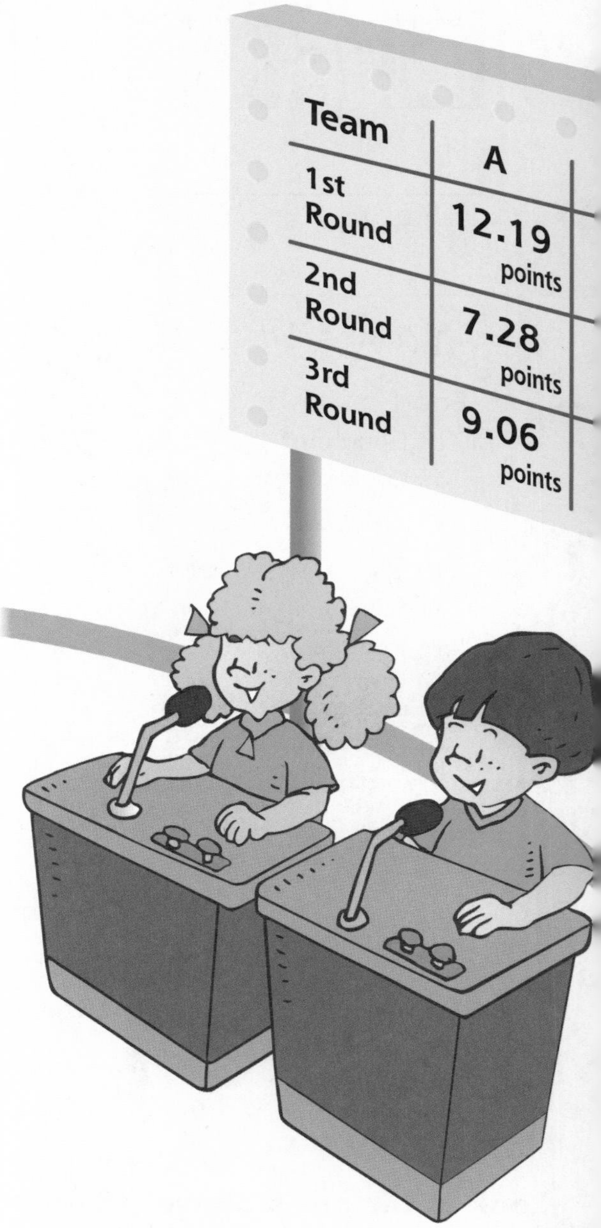
Team
A
1st Round
12.19 points
2nd Round
7.28 points
3rd Round
9.06 points

See how many points each team has got in the Math Contest. Help the children solve the problems. (6 marks)

(54) How many points has Team A got in all?

Team A has got ________ points in all.

(55) Team B has got 4.9 points fewer than Team A. How many points has Team B got?

Team B has got ________ points.

(56) Each team has to answer 20 questions. If Team A answers 15 questions correctly, which circle graph shows the information properly? Check ✔ the letter.

A
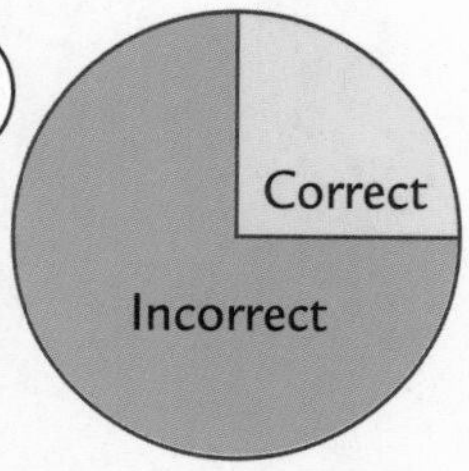
Correct
Incorrect

B
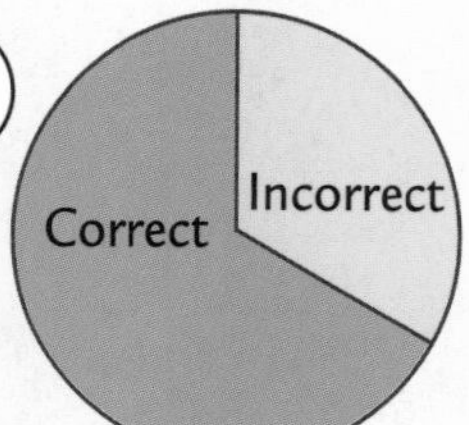
Correct
Incorrect

C
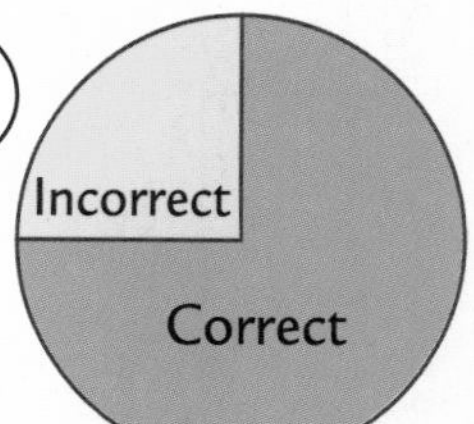
Incorrect
Correct

Score

100

ISBN: 978-1-927042-13-7

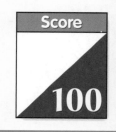

1 Numbers to 10 000

1. 3456 ; 3000 + 400 + 50 + 6 ;
   Three thousand four hundred fifty-six
2. 4235 ; 4000 + 200 + 30 + 5 ;
   Four thousand two hundred thirty-five
3. Four thousand five
4. Three thousand two hundred
5. One thousand sixty
6. Five thousand two hundred seventy-four
7. Nine thousand eight hundred sixty-three
8. Seven thousand forty-nine
9. Six thousand seven hundred fifty-one
10. 5276   11. 3482   12. 2309
13. 7540   14. 300   15. 2000
16. 80   17. 9
18. <   19. <   20. >
21. <   22. >
23. 4802 > 2867 > 2308 > 1988
24. 6342 > 4632 > 3264 > 2364
25. 7005 > 5700 > 5070 > 5007
26. 6600 ; 6400 ; 6300   27. 8040 ; 8050 ; 8070
28. 7444 ; 5444 ; 4444   29. 6000 ; 6002 ; 6003
30. 1459   31. 3047   32. 2004
33. I   34. I   35. III
36. 2004, 2040, 2400, 4002, 4020, 4200

Activity

1. 4444   2. 5555
3. 66   4. 77

2 Addition and Subtraction

1. 1488   2. 8112   3. 6010
4. 6759   5. 9343   6. 7385
7. 7573   8. 9011   9. 6654
10. 6742
11.

12. 5811   13. 9452   14. 2110
15. 4123   16. 2338   17. 7276
18. 8160   19. 5891   20. 6185
21. 4445
22. ✔ ;
$$\begin{array}{r} 865 \\ +\ 2135 \\ \hline 3000 \end{array}$$
23. ✗ ;
$$\begin{array}{r} 329 \\ +\ 4696 \\ \hline 5025 \end{array}$$
24. ✗ ;
$$\begin{array}{r} 977 \\ +\ 7044 \\ \hline 8021 \end{array}$$
25. ✔ ;
$$\begin{array}{r} 683 \\ +\ 6363 \\ \hline 7046 \end{array}$$

26. 1500   27. 6800   28. 5200
29. 1100   30. 7100   31. 9900
32. 3100   33. 1200   34. 9000
35. 3000   36. 4000   37. 4000
38. 9000   39. 8000
40. 7920 ;
$$\begin{array}{r} 4200 \\ +\ 3700 \\ \hline 7900 \end{array}$$
41. 8051 ;
$$\begin{array}{r} 9000 \\ -\ 900 \\ \hline 8100 \end{array}$$
42. 6176 ;
$$\begin{array}{r} 7100 \\ -\ 900 \\ \hline 6200 \end{array}$$
43. 7480 ;
$$\begin{array}{r} 5600 \\ +\ 1900 \\ \hline 7500 \end{array}$$
44. 8429 ; 9100 − 700 = 8400
45. 6151 ; 3500 + 2700 = 6200
46. 1258 + 1258 ; 2516 ; 2516
47. 2016 − 986 ; 1030 ; 1030
48. 5000 − 817 ; 4183 ; 4183

Activity

1.
$$\begin{array}{r} 1\ \boxed{2}\ 0\ 7 \\ +\ 1\ 9\ \boxed{5} \\ \hline 1\ 4\ 0\ 2 \end{array}$$
2.
$$\begin{array}{r} 3\ 2\ \boxed{1}\ 1 \\ -\ \boxed{6}\ 0\ 9 \\ \hline 2\ 6\ 0\ 2 \end{array}$$
3.
$$\begin{array}{r} 2\ 7\ 3\ \boxed{4} \\ -1\ 8\ \boxed{7}\ 5 \\ \hline 8\ 5\ 9 \end{array}$$

3 Multiplication

1. 50   2. 90   3. 80
4. 100   5. 700   6. 200
7. 3000   8. 4000
9. 2 ; 8 ; 800   10. 3 ; 6 ; 60
11. 2000   12. 2100   13. 18 000
14. 240   15. 12 000   16. 720
17. 4500   18. 18 000   19. 26
20. 96   21. 54   22. 160
23. 588   24. 79   25. 126
26. 292   27. 245   28. 208
29. 166   30. 259
31a. 185   b. 296
32a. 120   b. 272
33. 6 ; 3 ; 5 ; 536   34. 639   35. 1500
36. 2595   37. 1098   38. 2466
39. 3264   40. 5684
41. 1260 ; 2940 ; 3360 ; 3780
42. 1300 ; 3250 ; 3900
43. 162 ;
$$\begin{array}{r} 18 \\ \times\ 9 \\ \hline 162 \end{array}$$
44. 2275 ;
$$\begin{array}{r} 325 \\ \times\ 7 \\ \hline 2275 \end{array}$$
45.
$$\begin{array}{r} 15 \\ \times\ 9 \\ \hline 135 \end{array}$$
$$\begin{array}{r} 13 \\ \times\ 8 \\ \hline 104 \end{array}$$
; 239

Activity

6

4 Division

1.
```
    1        1        1 2 R 2
6)7 4    6)7 4    6)7 4
  6        6          6
  1        1 4       1 4
                     1 2
                       2
```
;   ;   ;

12 ; 2

2.
```
  2 3
3)6 9
  6
  9
  9
```

3.
```
  1 3
4)5 2
  4
  1 2
  1 2
```

4.
```
  1 3 R 5
7)9 6
  7
  2 6
  2 1
    5
```

5.
```
  1 6 R 4
5)8 4
  5
  3 4
  3 0
    4
```

6. 13   7. 19R1   8. 15R1
9. 9R1   10. 8R1   11. 8R3

12.
```
    1          1 4         1 4 8 R 1
3)4 4 5    3)4 4 5     3)4 4 5
  3          3            3
  1          1 4          1 4
             1 2          1 2
               2          2 5
                          2 4
                            1
```
;   ;   ;

148R1

13.
```
  1 2 1
4)4 8 4
  4
  8
  8
  4
  4
```

14.
```
  1 2 6
7)8 8 2
  7
  1 8
  1 4
  4 2
  4 2
```

15.
```
  1 1 9 R 1
6)7 1 5
  6
  1 1
    6
  5 5
  5 4
    1
```

16.
```
  1 0 5 R 8
9)9 5 3
  9
  5 3
  4 5
    8
```

17. 212   18. 62R5   19. 84
20. 23R4   21. 72   22. 29 ; 3

23.
```
  1 4
5)7 2
  5
  2 2
  2 0
    2
```

24. ✗
```
  2 7 R 1
3)8 2    3)8 2
  3        6
  5 2      2 2
           2 1
             1
```

25.
```
  1 2       1 2 R 2
3)3 8    3)3 8
  3        3
  8        8
  6        6
  ✗        2
```

26.
```
  1 3       1 3 R 2
6)8 0    6)8 0
  6        6
  2 0      2 0
  1 ✗      1 8
             2
```

27. ✔ ; 18 x 5 = 90
28. 37 ; 38 x 2 = 76
29. 190 ; 180 x 4 = 720
30. ✔ ; 123 x 7 = 861
31. 105 ÷ 5 ; 21 ; 21
32. 168 ÷ 7 ; 24 ; 24
33. 100 ÷ 7 ; 14R2 ; 14
34. 197 ÷ 6 ; 32R5 ; 33

Activity

18

5 Time

1. 50   2. 70
3. Internet   4. vacuum cleaner
5. electric guitar   6. dryer
7. 50   8. 30
9. 20   10. 400
11. 8   12. 7000
13. 42 ; 28 ; 22 ; 28   14. Tony
15. 3   16. 14 min

Activity

Kevin

6 Perimeter and Area

1.

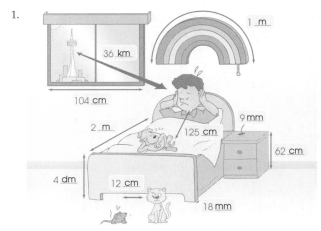

2.

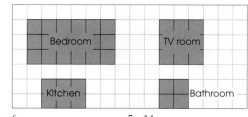

3. (Suggested answer)

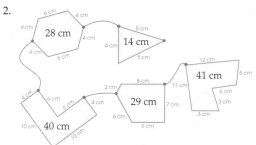

4. 6   5. 14
6. 10   7. 16

8. (Suggested answer)

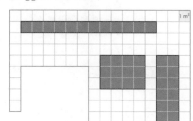

9. Perimeter: 12 ; 14 ; 16 ; 16

Area: 6 ; 9 ; 10 ; 14

10. D   11. A

12a. C and D   b. No

Activity

C

7   2-D Shapes

1.

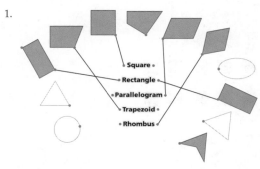

2. A : Rectangle   B : Rhombus

C : Trapezoid   D : Square

E : Parallelogram

3. A, B, D, and E   4. B, C, and E

5. A and D

6-7. (Suggested answers)

6. 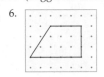   7.

8. congruent   9. similar

10. similar   11. congruent

12. similar   13. congruent

14.

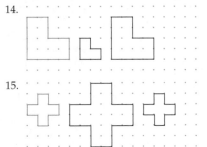

15.

16.

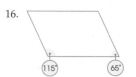

17.

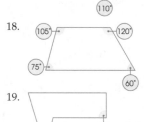

18.

19.

Activity

(Suggested answer)

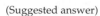

8   3-D Figures

1. Rectangular prism ; ⬜ ; ⬜ ; ⬜

2. Rectangular pyramid ; △ ; ⊠ ; △

3. Hexagonal prism ; ⬜ ; ⬡ ; ⬜

4. ; hexagonal prism ; 18 ; 12 ; 8

5. ; pentagonal pyramid ; 10 ; 6 ; 6

6. ; triangular pyramid ; 6 ; 4 ; 4 ; triangle

7. ; rectangular prism ; 12 ; 8 ; 6 ; rectangle

8. ; rectangular pyramid ; 8 ; 5 ; 5 ; 4

9. ; square ; cube

10. 12 ; 18   11. 4 ; 6   12. 8 ; 12

13. 5 ; 8   14. 10 ; 15   15. 7 ; 12

16. Rectangular prism

17. Rectangular prism or hexagonal pyramid

18. 6 ; 10   19. 14 ; 21
20.   21.

22. (triangular pyramid figure)   23. (hexagonal prism figure)

Activity

1. 36 ; 19   2. 30 ; 20

Midway Test

1. 20   2. 20   3. 80
4. 2   5. 40
6. Five thousand four hundred ten
7. Seven thousand eight hundred ninety-nine
8. Nine thousand three hundred one
9. A : 68°   B : 52°   C : 60°
10. B, C, A
11. 2229   12. 9362   13. 2121
14. 1775   15. 6122   16. 6617
17. 8094
18. ✔ ; 167 + 6146 = 6313
19. 4445 ; 555 + 4555 = 5110
20. Rectangle ; 16 cm   21. Parallelogram ; 18 cm
22. Congruent pairs: N, D and K, E and M, J and L
    Similar pairs: A and G, B and E, B and M, D and I, G and N, I
                and K, J and O, L and O
23. Pentagonal pyramid   24. Cube
25. Rectangular pyramid
26. 600 ; 1   27. 8000 ; 9
28. 5515   29. 595   30. 460
31. 584   32. 897   33. 750
34. Triangular prism ; Rectangular prism
35. 9   36. 5 ; 8   37. 24
38. 24   39. 13R2   40. 137
41. 116R2   42. 15R2   43. 233R2
44. 1728 + 1697 = 3425 ; 3425
45. 1697 − 1198 = 499 ; 499
46. 9840 ÷ 5 = 1968 ; 1968
47. 4 x 57 = 228 ; 228

9   *Fractions*

1a. $\frac{5}{11}$ b. $\frac{4}{11}$ c. $\frac{2}{11}$
2. $\frac{4}{7}, \frac{3}{8}, \frac{1}{5}, \frac{5}{14}$ 3. $\frac{9}{2}, \frac{11}{3}, \frac{13}{6}, \frac{2}{1}$
4. $3\frac{1}{6}, 1\frac{1}{9}, 2\frac{5}{7}, 1\frac{4}{5}$
5. $2\frac{1}{4}$ 6. $3\frac{2}{3}$
7. $5\frac{2}{6}$ 8. $3\frac{3}{6}$
9. $2\frac{1}{5}$ 10. $1\frac{1}{5}$ 11. $2\frac{3}{5}$

12. (fraction grid)   13. (fraction grid)
14. <   15. >   16. <
17. >   18. <   19. <
20. <   21. >   22. <
23. <   24. >   25. <
26. (circles) ; $\frac{5}{6}, \frac{3}{6}, \frac{1}{6}$
27. (grids) ; $1\frac{1}{3}, 1\frac{2}{3}, 2\frac{1}{3}$
28. $\frac{2}{7}, \frac{3}{7}, \frac{6}{7}$ 29. $5\frac{1}{6}, 5\frac{3}{6}, 6\frac{5}{6}$
30. (flowers) ; $\frac{3}{12}$
31. (kites) ; $\frac{3}{10}$

Activity
(faces) ; 2

10   *Decimals*

1. $\frac{4}{10}$; 0.4   2. $2\frac{6}{10}$; 2.6
3. $\frac{37}{100}$; 0.37   4. $2\frac{9}{100}$; 2.09
5. 0.3   6. 0.51   7. 0.02
8. 1.9   9. $1\frac{69}{100}$ 10. $\frac{8}{10}$
11. $5\frac{3}{100}$ 12. $4\frac{6}{10}$ 13. 0.5
14. 0.20   15. 0.03   16. 1.7
17. 0.61   18. 2.12
19. Nine tenths
20. Fifty-eight hundredths
21. One and four hundredths
22. Seven and twenty-three hundredths
23. Tenths ; 0.7
24. Hundredths ; 0.09
25. Ones ; 3
26. Hundredths ; 0.01
27. 3.40, 3.4   28. 0.25, 0.250
29. 0.6, 0.60   30. 0.08, 0.080

31. $24.9

32. $2.39

33. $7.99

34. $0.5

35. $17.95

36. 2.86, 3.4, 3.84, 3.98   37. 3.84, 3.98, 4.7

38. 2.86 and 3.84

39. 2.86, 3.4, 3.84, 3.98, 4.7, 6.6, 7.15, 8.14

40. 1.09   41. 1.12   42. 0.98

43. 1.03   44. 1.05   45. 1.08

46. Aaron   47. Fiona

48. Fiona ; Jerry ; Nancy ; Linda ; Jimmy ; Aaron

Activity

1. 0.12   2. 0.34

3. 0.25   4. 0.29

11 Addition and Subtraction with Decimals

1. 0.4 ; 0.3 ; 0.7   2. 0.9 ; 0.5 ; 0.4

3. 1.2 ; 1.6 ; 2.8   4. 2.5 ; 1.3 ; 1.2

5. 1.9 ; 1.3 ; 3.2

6.
2nd Add the hundredths.	3rd Add the tenths.	4th Add the ones.
1 . 3 2 + 0 . 5 9 [1]	1 . 3 2 + 0 . 5 9 [9] 1	1 . 3 2 + 0 . 5 9 [1] . 9 1

7.
2nd Subtract the hundredths.	3rd Subtract the tenths.	4th Subtract the ones.
1 . 5 6 − 0 . 4 8 [8]	1 . 5 6 − 0 . 4 8 [0] 8	1 . 5 6 − 0 . 4 8 [1] . 0 8

8. 0.90   9. 10.03   10. 9.08

11. 9.05   12. 1.88   13. 0.89

14. 4.62   15. 11.05   16. 7.92

17. 11.23   18. 8   19. 2.22

20. 5.9   21. 0.38

22. 4 ; 2 + 2 = 4   23. 4.77 ; 6 − 1 = 5

24. 9.56 ; 8 + 1 = 9   25. 7.61 ; 9 − 1 = 8

26. 2.41 ; 1 + 2 = 3

27. 78.76   28. 48.16

29. 8.94   30. 18.08 ; 16.62 ; 18.22

31. Beauty   32. 20 − 18.08 ; 1.92 ; 1.92

33. 9.24 − 7.68 ; 1.56 ; 1.56

Activity

1. A   2. C   3. B

12 Money

1-2. (Individual estimates)

1. 33.32   2. 42.26

3. 10 ; 6.32   4. 15 ; 21.50

5. 5 ; 4.82   6. 20 ; 30.84

7. 10.12 ; 13.47 ; 15.45 ; 42.77

8. 10.12 + 13.47 ; 23.59 ; 23.59

9. 50 − 42.77 ; 7.23 ; 7.23

10. 15.45 + 15.45 ; 30.90 ; 30.90

Activity

13 Capacity, Volume, and Mass

1. L   2. mL   3. L

4. mL   5. L   6. L

7. 5000   8. 7   9. 9

10. 3000   11. 1020   12. 5400

13. 2008   14. 3086

15. 8 ; 500 ; 1700 ; 850

16. 3   17. 250 mL   18. 1 L

19. A : 8   B : 10

C : 28   D : 34

E : 22   F : 15

20. (Suggested answer)

21. 4   22. 3   23. 9 ; 14 ; 19

24. C   25. A   26. 5

27. 130   28. 8500

Activity

A

14 Patterning

1.   2.

3.

4. Orientation, Colour   5. Shape, Size

6. 25 ; 41   7. 30 ; 36

8. 78 ; 66   9. 45 ; 75 ; 120

10. 27 ; 35 ; 54

11. 46 ; 56 ; Subtract 8, add 10

12. 46 ; 43 ; Multiplied by 2, subtract 3

13. 64 ; 32 ; Multiplied by 4, divided by 2

14. 15 + 9 = 6 + 18   15. 29 − 6 = 20 + 3

15 + 10 = 6 + 19   29 − 7 = 20 + 2

15 + 11 = 6 + 20   29 − 8 = 20 + 1

16.   17.   18.

19a. increases ; 3   b. 15 ; 5 ; 10
20a. increases ; 2   b. 9
21a. decreases ; 2   b. 5th

22.

No. of Children	1	2	3	4	5	6
Total No. of 🍎	6	12	18	24	30	36
Total No. of 🍎	9	18	27	36	45	54
Total No. of Apples	15	30	45	60	75	90

23. 6   24. 63   25. 120

26.

Activity

75

15 Transformation and Coordinates

1.

2.

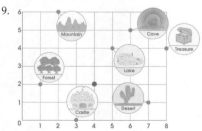

3.

4. Castle   5. Cave
6. (2,6)   7. (8,4)
8. 5 ; down ; 3 ; left
9.

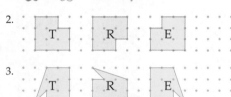

Activity

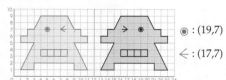

◉ : (19,7)
⇐ : (17,7)

16 Graphs and Probability

1.

Bacon	ⵌ ‖
Ham	‖‖
Eggs	ⵌ
Sausages	ⵌ ⵌ

2. Eggs   3. Ham
4. 20   5. 23
6.

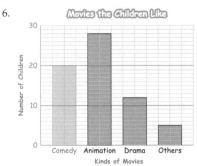

Movies the Children Like

7. 28   8. 8   9. 65
10. Number of Children   11. Kinds of Movies
12.

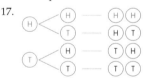

13. "Others" refers to movies that are not specified on this graph, e.g. horror movies.
14. equally probable   15. less probable
16. more probable
17.

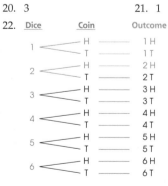

18. 4   19. No
20. 3   21. 1
22.

Dice	Coin	Outcome
1	H / T	1 H / 1 T
2	H / T	2 H / 2 T
3	H / T	3 H / 3 T
4	H / T	4 H / 4 T
5	H / T	5 H / 5 T
6	H / T	6 H / 6 T

23. 12   24. 6
25. 4   26. No
27. Each outcome will occur about 10 times because all the outcomes are equally probable. The number of times of getting each outcome should be about the same.

Activity

Yes, because no matter who plays the game, the probability of getting each outcome is the same.

Final Test

1. Green: $\frac{8}{9}$; $\frac{10}{11}$ Red: $\frac{5}{4}$; $\frac{9}{7}$

   Blue: $2\frac{1}{3}$; $4\frac{9}{10}$

2. 0.7 3. $\frac{3}{100}$ 4. $1\frac{11}{100}$

5. 2.29 6. $4\frac{7}{10}$ 7. 8.9

8. 0.7 9. 0.03 10. 10

11. 4

12. [leaf] [leaf] ; Shape ; Colour

13. $\frac{7}{8}$ 14. $4\frac{3}{7}$ 15. 4.50

16. 3.09 17. 19.03 18. 7.45

19. 6.58 20. 1.33 21. Less than

22. 13.16

23a. 4 b. 16

24a. 13 b. 30

25.

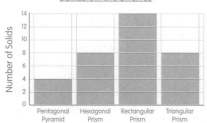

Solids on the Shelves

26. 8 27. 14

28a. (2,6) b. (8,0)

  c. (1,1) d. (5,4)

29. (Suggested answer)

30. 6 31. (3,3) 32. 992

33. 119R3 34. 2385 35. 271R2

36. more probable 37. less probable

38. equally probable 39. 35 ; 42 ; 49 ; 56 ; 63

40. 56 41. 10

42. 4 43. Translation

44. Rotation 45. Reflection

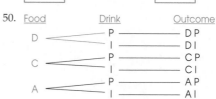

46. [flower] 47. [flower]

48. [flower] 49. [cherry]

50.
Food	Drink	Outcome
D	P	D P
	I	D I
C	P	C P
	I	C I
A	P	A P
	I	A I

51. 4.87 ; 4 + 1 = 5 52. 5.08 ; 1 + 4 = 5

53. 16.58 ; 9 + 8 = 17

54. 12.19 + 7.28 + 9.06 = 28.53 ; 28.53 .

55. 28.53 − 4.9 = 23.63 ; 23.63

56. C

ISBN: 978-1-927042-13-7